Why
Grow
Here

Diane. Please enjoy!

Kathryn Chase Merrett

KATHRYN CHASE MERRETT

Why Grow Here

Essays on Edmonton's Gardening History

᙮ The University of Alberta Press

Published by

The University of Alberta Press
Ring House 2
Edmonton, Alberta, Canada T6G 2E1
www.uap.ualberta.ca

Copyright © 2015 Kathryn Chase Merrett

LIBRARY AND ARCHIVES CANADA
CATALOGUING IN PUBLICATION

Merrett, Kathryn Chase, 1944–, author
    Why grow here : essays on Edmonton's gardening history / Kathryn Chase Merrett.

Includes bibliographical references and index.
Issued in print and electronic formats.
ISBN 978-1-77212-048-6 (paperback).—
ISBN 978-1-77212-075-2 (EPUB).—
ISBN 978-1-77212-076-9 (kindle).—
ISBN 978-1-77212-077-6 (PDF)

    1. Gardening—Alberta—Edmonton—History. I. Title.

SB453.3.C2M47 2015   635.097123'34
C2015-902530-3
C2015-902531-1

Index available in print and PDF editions.

First edition, first printing, 2015.
Printed and bound in Canada by Houghton Boston Printers, Saskatoon, Saskatchewan.
Copyediting and proofreading by Kirsten Craven.
Indexing by Judy Dunlop.

All rights reserved. No part of this publication may be reproduced, stored in a retrieval system, or transmitted in any form or by any means (electronic, mechanical, photocopying, recording, or otherwise) without prior written consent. Contact the University of Alberta Press for further details.

The University of Alberta Press is committed to protecting our natural environment. As part of our efforts, this book is printed on Enviro Paper: it contains 100% post-consumer recycled fibres and is acid- and chlorine-free.

The University of Alberta Press gratefully acknowledges the support received for its publishing program from The Canada Council for the Arts. The University of Alberta Press also gratefully acknowledges the financial support of the Government of Canada through the Canada Book Fund (CBF) and the Government of Alberta through the Alberta Media Fund (AMF) for its publishing activities.

For Rob, at heart a gardener.

ZONES OF HARDINESS

A visitor from down south stared at my apple tree
and said: "Those don't grow here you know. It's too cold."
If the apricot tree in Highlands knew it couldn't live here,
it might stop scattering white blossoms over three lawns.
Daffodil bulbs that we planted in old wash tubs
came up in the dark garage this March,
raising their yellow flags in a delusion of spring.
We were too tactful to correct them.
But the visitor, the visitor got apple pie for dessert.

—BERT ALMON, 2004

< Tom Daly's apple tree at Clover Bar, east of Edmonton, c. 1905. [PAA B.9044]

Preface xi

Acknowledgements xvii

Abbreviations xix

Contents

1 | Why Grow Here? 1

2 | Donald Ross 19
   Edmonton's "Father of Gardening"

3 | Nature's Garden Transformed 37

4 | Among the First 65

5 | The Edmonton Horticultural Society 101
   Working for the City Beautiful

6 | Waste Places 139
   Vacant Lot Gardening in Edmonton

7 | Edmonton, the Rose City 167

8 | The Invisible Tapestry 197
   Remembering Edmonton's Chinese Gardeners

9 | Citizen Gardeners 241

Notes 259

Sources 297

Index 307

Preface

QUESTIONS ABOUT EDMONTON'S GARDENING HISTORY began to form in my mind years ago while researching a book on the Edmonton City Market. I discovered, for instance, that in the 1940s and 1950s, the Edmonton Horticultural Society had staged horticultural shows in Edmonton's first covered market building, although my questions about the society—how long it had been in operation, and what it did—remained, until much later, unanswered.[1] At that time the society's records were being stored in someone's basement and were not accessible except for one short, supervised search for a specific item. At some point during my research, I came across references to a Chinese market gardener who claimed to have been refused a stall at the market on account of his ethnicity. Who was this gardener and where were the other Chinese market gardeners who were said to have cultivated plots in the North Saskatchewan River valley in the first half of the twentieth century? When I discovered that a few of Edmonton's early market vendors had been interested in plant breeding and that one of them, Robert Simonet, had become famous for his double-flowering petunias and recognized nationally for his work in plant hybridization, I was curious. What was it about Edmonton that turned gardeners into plant breeders? And why were Edmonton's early gardeners, including many of the market vendors I interviewed, so optimistic about what could be grown here, despite long, cold winters, short summers, wind, dryness, and rapid changes in temperature?

Gardeners complain about the weather today; some leave the city for more temperate places where they can garden year-round. And yet, far from being intimidated by the climate, Edmonton's early settlers happily gave themselves over to exploring the scope for growing here. Some of these, cheered by the generally high quality of the soil, buoyed up by the effects of long summer days on certain plants and confident they could apply their skills to develop better cultivation techniques, hardier plants, and more beautiful gardens, set to work to make miracles. Horticultural optimism has been an important feature of Edmonton's gardening history, and the first essay in this book moves backwards and forwards through Edmonton's settlement history to source the roots of that optimism.

Donald Ross, who counts as one of Edmonton's very first settlers, is generally remembered in one or another of the many contexts in which he made his mark on Edmonton: coal mining, hotel keeping, real estate, even pork packing. In the second essay in this collection, however, he makes an appearance as a gardener and the argument is advanced that if anyone deserves to be acknowledged as the gardening elder of the post-Hudson's Bay period in Edmonton's short history, it is Ross.

"Nature's Garden Transformed" deals with conflicting definitions of the word "garden," and evolving definitions of the word "nature," as much as these can be inferred from a variety of local historical and fictional sources. It takes the reader on a highly selective trip from Edmonton's presettlement period to today's interest in complex and elusive concepts of ecology.

It would be impossible to introduce all the many characters who have helped to shape, in one way or another, Edmonton's gardening history. The four whose stories are briefly told in the fourth essay—florist Walter Ramsay; seedsman Alfred Pike; and plant breeders Georges Bugnet and Robert Simonet—were among the first in Edmonton to step away from a course that interested them less in order to embark on a horticultural project that interested them more.

Some cities acquire reputations as beautiful places to live. Others aspire to such reputations and fall short of the mark. The fifth essay traces efforts made by the Edmonton Horticultural Society over a period of more than a century to ensure that Edmonton would be securely located in the first category.

"Waste Places" begins with the formation of a Vacant Lots Garden Club in Edmonton during the First World War but goes on to show how the informing ideology of vacant lot gardening has changed over time, leading to the demise of the vacant lots program in 1989 and the emergence of the community gardening movement in the late twentieth and twenty-first centuries.

The seventh essay is all about roses, and the fascination Edmontonians have had for the "queen of flowers" in spite of the city's reputation for frigid winters and windy, dry summers. Perversely, however, climate and weather have proved to be an inspiration rather than a disincentive; history confirms that efforts made in the 1930s and 1940s to brand Edmonton as the "Rose City" may not have been as unrealizable as one might think.

In 2004, I remember being surprised to hear that the Chinese community in Edmonton planned to build a "traditional" Chinese garden to commemorate the role played by the Chinese in the city's history. "Why a traditional garden?" I wondered, when early Edmontonians from China had contributed so much to the city through their accomplishments as market gardeners. "The Invisible Tapestry" attempts to put names and faces to some of these market gardeners and to suggest how their work as gardeners fits into a broader cultural context.

"Citizen Gardeners," the last essay in the collection, profiles three very different individuals whose contributions to the city have been mediated through the garden—photographer and gardening advocate Gladys Reeves; gardener and public figure Lois Hole; horticulturist and public servant John Helder. These are three very different individuals who have made their mark at different times in Edmonton's short history and in different ways; what they share is the notion that the garden is somehow a model for, and an indicator of, the health of a civil society.

Each of the essays in this collection is written to stand alone. The sources for each one vary and include archival records, records supplied or loaned by companies and organizations, newspapers, fiction, and oral interviews. Because of the thematic nature of the essays, and their tendency to move backwards and forwards in time, gardeners and gardening advocates who figure in one essay are likely to figure in others. Repeat appearances by figures who were active in several spheres has been adopted as a compositional strategy, although I have tried to minimize the repetition of too many biographical details. The first time an individual appears in a given essay his or her name is given in full; subsequent references in that essay are shortened to the last name.

Certain essays in the collection may have particular reader constituencies, including the eighth, on Chinese market gardeners, and the fifth, which focuses on the work of the Edmonton Horticultural Society. I hope, however, that for those who choose to read the essays sequentially, the whole will add up to something more than the sum of its parts.

In two areas, nomenclature is an issue and I need to explain the systems I have adopted; plant names and references to Aboriginals or Aboriginal groups. In the rare instances where I have referred to plants by their botanical names, I have adopted the convention of capitalizing the genus but not the species and italicizing both words (*Rosa acicularis*). Common names of plants such as elm, prickly rose, and dandelion are not capitalized or italicized. Named hybrids are capitalized and enclosed in single quotes ('Earl Haig' rose).

In deciding how to deal with the sensitive issue of nomenclature as it applies to status Indians and to Métis, I have referred to the University of British Columbia's Indigenous Foundations website.[2] In accordance with this site, I have relied quite heavily on the term "Aboriginal" to refer to any descendant of Canada's first inhabitants, including First Nations and Métis peoples. Where I can be more specific by referring to either the First Nation to which an individual belonged or the band from which he or she came, I do. Where the term "Indian" or "Native" exists in a quotation, I have left it. Where a direct quotation betrays racist views on the part of the speaker, or where the speaker is engaging in racial stereotyping, again I have left these quotes intact.

I expect the majority of this book's readers to be Edmontonians. Some will be gardeners. Others will be aficionados of Edmonton's local history. A few may be interested in the thing that interests me most, discovering what it is about the history of any given city that makes it, in some way, unique. In the interests of readability, I have incorporated as many references as possible into the text. Notes and occasional elaborations on the text are aggregated at the end of the book under their various essay titles. There is also a brief list of works cited and consulted.

Finally, I am certain that anyone reading this book will recognize absences from its pages, individuals who should have been mentioned or topics that could have been treated. The history of the Devonian Botanic Garden and the many constituent gardens that go to make it up could be a book in itself. The history of the Department of Horticulture in the Faculty of Agriculture at the University of Alberta, including the research carried out there and the outreach offered for years by its professors and students to home gardeners, much of it through the Faculty of Extension, is an integral part of the social and cultural history of gardening in Edmonton and it is not treated here. The many gardening businesses that were started by an individual or a couple and went on to become large, family-run operations would make an interesting composite history. Edmonton and area's many lily breeders, including the schoolteacher Fred Tarlton, who registered a host of beautiful martagon lilies and inspired others to follow in his footsteps, deserve attention that is missing in this book. All I can say is that I am more than aware of these omissions and hope others will continue where I have left off.

Acknowledgements

RESEARCH FOR THIS BOOK began in 2002 with a grant from the Alberta Historical Resources Foundation to study the social and cultural history of gardening in Edmonton. In addition to the financial incentive such grants offer their recipients, they provide a certain moral incentive to make the results of the research available to the public. I am grateful to the foundation for both sorts of incentive and for providing additional publication funding through a grant by the Alberta Lottery Fund. I would also like to thank the Edmonton and District Historical Society and the Alberta Horticultural Association for asking me, in 2006 and 2011, respectively, to speak to their organizations about the history of horticulture in the Edmonton area. These talks gave me the idea for the theme that runs through the book, that Edmonton's gardening culture has developed as a direct response to the challenges imposed by climate and geography.

Much of my research has been carried out in libraries and archives where I have been unfailingly met with helpful and courteous service. These include the Provincial Archives of Alberta; the Hudson's Bay Company Archives, a division of the Archives of Manitoba; the Glenbow Archives in Calgary; the Bruce Peel Special Collections Library at the

University of Alberta; the Alberta Government Library and reference librarian Connie Hruday; the library at the University of Alberta Devonian Botanic Garden and Gordon Nielson, who facilitated access to it; and especially the City of Edmonton Archives where Paula Aurini-Onderwater, Sharon Bell, Lorraine Butchart, Kim Christie-Milley, and Tim O'Grady all made special efforts on my behalf.

Before beginning the research for this book, I had no particular interest in gardening and no formal training in horticulture. My two first interviewees, Mary Shewchuk and Lorrie McFadden, volunteered their expertise at a meeting of the Edmonton Horticultural Society, probably some time in 2002. They were my happy initiation to the world of gardening and gardeners; their eagerness to share what they knew was more than encouraging but perhaps I need not have worried. Gardeners and horticulturists love to share what they know and do and accomplish. With the help of many people who took the time to talk to me, my eyes were opened not just to the pleasures and beauties of the garden but to the role that gardening plays in the life of any place. This book could not have been written without the help of gardeners.

Bert Almon's poem, *Zones of Hardiness*, has been an inspiration since I first read it in 2005; I always knew it would be the epigraph.

Several people read all or portions of the text and I have benefitted from their comments and suggestions, even when I have stubbornly resisted some of them. Rob Merrett listened to the entire text read aloud, reacting enthusiastically to the content even as he caught grammatical errors and lexical imprecisions.

I want to thank the anonymous readers who read the book for the University of Alberta Press and made many helpful comments and recommendations; my editor Kirsten Craven who helped take the terror out of the meticulous work of copyediting; and Judy Dunlop, indexer par excellence whose services I am lucky to have received.

And finally, my thanks to Linda Cameron and the talented crew she has assembled at the University of Alberta Press, senior editor Peter Midgley, designer Alan Brownoff, production editor Mary Lou Roy, and the wonderful marketing team of Cathie Crooks and Monika Igali. The entire team displays an admirable love of the book and, what is even better for authors, a love of every book the press publishes.

For any mistakes in the conception or execution of this project, I take full responsibility.

Abbreviations

ABBREVIATIONS are used mainly in the notes. When used in an essay, the first reference is given in full with the abbreviation in brackets.

The first references in each essay to either the *Edmonton Bulletin* or the *Edmonton Journal* are given in full. Subsequent references to these two newspapers in each chapter are shortened to "the *Bulletin*" and "the *Journal*," respectively. Although the *Edmonton Bulletin* began in December 1880 simply as the *Bulletin*, and was published from time to time under titles such as the *Daily Edmonton Bulletin*, the *Edmonton Daily Bulletin*, and the *Morning Bulletin*, I have used only the one name.

Following is a list of abbreviations:

CEA	City of Edmonton Archives
EB	*Edmonton Bulletin* (the *Bulletin*)
EJ	*Edmonton Journal* (the *Journal*)
GA	Glenbow Archives
HBC	Hudson's Bay Company
HBCA	Hudson's Bay Company Archives, Archives of Manitoba
PAA	Provincial Archives of Alberta
UAA	University of Alberta Archives

1 Why Grow Here?

When I was twelve years of age, my father's health broke [and] the doctors ordered a sea voyage; we came to Edmonton, [and] within a month of our arrival my Dad was preparing his garden. Seeing him chopping down trees Old Donald Ross...asked what he was going to do. "Start a garden" he said, "Oh" replied Mr. Ross "nothing but taters [and] cabbage will grow here."

—GLADYS REEVES

THE RELATED FIELDS of agriculture and horticulture assumed a new importance in Edmonton in 1870, the year Rupert's Land was sold to the Canadian government by the Hudson's Bay Company, because the prospect of permanence for a fledgling settlement that had grown up around the fur trade brought with it an imperative to attract settlers. For almost a hundred years before 1870, fur traders and their families had been content to live with successes and failures in the company's gardens and fields. Were these successes and failures to be taken as givens? It would seem that Edmonton's first settlers and gardeners made a leap of faith when they set out to prove that Edmonton would become, as a traveller predicted in 1883, the "future garden of the North-West."[1] Why did they choose to make this leap? And how did they prove to themselves and to others that they had not made a dreadful mistake?

Frank Oliver, founder of the *Edmonton Bulletin* in 1880 and then Minister of the Interior in the government of Sir Wilfrid Laurier, was so convinced he had not mistaken his town's gardening potential that he became, through the pages of his newspaper, the most articulate spokesman for its future as an agricultural and horticultural Mecca. Because the *Bulletin* is so full of gardening news during the first decades of its existence, it offers insights into the horticultural aspirations and achievements of Edmonton's first settlers.

By 1892, the year Edmonton was incorporated as a town, its citizens had firmly grasped the significance of gardening to the town's future. They knew that potential settlers from eastern Canada perceived Edmonton's northerly location as a disadvantage. An article published in a supplement to the March 6, 1892, edition of the *Bulletin* articulated the so-called eastern prejudice that only horticultural and agricultural successes could combat. "The great drawback to the general desirability of the Northwest in eastern eyes," it said, "is the supposition that it is entirely a treeless waste, with a uniform winter climate of long continued extreme cold and frequent severe storms."[2]

The eastern prejudice lay behind the creation of the Edmonton Agricultural Society in 1879, "the first [such society] ever organized in the North-West," Oliver boasted.[3] The newly formed society held its first annual exhibition on October 15 that year for which Chief Factor Richard Hardisty offered two rooms in the fort, "one of which was devoted to the vegetables and grain and the other to ladies work." Although the society floundered for a few years afterward, reorganizing in 1882 to launch a second exhibition, Oliver immediately grasped its significance to the new settlement. Writing as Edmonton correspondent for the *Saskatchewan Herald*, he argued "that there is no better way

Vegetable display created for a visit to Edmonton by Lord Aberdeen, Governor General of Canada, and his wife Lady Aberdeen, 1894. [PAA B.9020]

to improve farm produce and show to the world the peculiar fertility of our soil, than by supporting an agricultural society," and it was an argument he continued to make as publisher of the *Bulletin*.[4]

In 1882, the *Bulletin* carried a detailed report of the second annual exhibition held on October 19 that year by the agricultural society. Not surprisingly perhaps, "[t]he show of field roots demonstrated fairly that in this particular branch Edmonton can not be surpassed, while that of garden vegetables proved the possibility of growing varieties which it has always been an article of faith to believe could not be grown."[5] In October 1886, roots and vegetables were claimed to have been "the strong point of the show," and if size mattered to prospective settlers, they surely would have been convinced:

> Cabbages girthed 4 ft 1 in, cauliflower 3 ft 1 ½ in, pumpkins perfectly
> ripe 4 ft 1 in, turnips 2 ft 4 in weighing 23 lbs., squash 4 ft 3 in,
> vegetable marrow 2 ft 1 in by 3 ft 7 in, beets 1 ft 8 in, potatoes 1 ft
> by 1 ft 8 in, white onions 1 ft 2 in, red onions 1 ft 1 ½ in, celery 3 ft
> 2 ½ in length, parsnips 3 ft 7 in, and beets 1 ft 4 ½ in. Every article
> in this class was faultless in shape, color and soundness, including
> the cucumbers, melons, citrons and tomatoes, the two latter being
> quite ripe.[6]

The flowers shown in the 1886 exhibition were "not numerous," but then nurseries and seed companies from Ontario and Winnipeg were just beginning to advertise a full range of flower seed in the *Bulletin* and it was a few years before floriculture was incorporated into the booster argument. In late September 1891, Edmonton sent many exhibits via the Canadian Pacific Railway for "exhibition in the east." In addition to grasses, grains, and supersized vegetables, a frame of honey was taken from Thomas Henderson's hives "and several bunches of flowers from the gardens in town, in proof not only of the capacity of this soil and climate for growing flowers but also in proof of the almost entire freedom from frost which we have enjoyed up to date."[7]

The *Bulletin* was Edmonton's first major vehicle for propaganda and its founder took advantage of any opportunity to promote the place through its garden produce. Residents who brought their newly dug potatoes to the *Bulletin* office to prove that they were bigger or better than those grown by someone else were always given a mention. Entire gardens were sometimes described in detail, especially when the description stood to alter readers' perceptions or provoke emulation:

> J. Knowles of Fraser Avenue, has locust trees two feet, apple trees a foot,
> catalpa six inches and Norway spruce an inch in height grown from
> seed and planted this spring, all making a healthy growth. Of 100 slips
> of golden willow received from the east only eight failed to grow, the
> remainder are growing vigorously, with shoots two feet in length. Wild
> goose berries transplanted into the garden this spring show a remarkable

vigorous growth having sent out shoots three and four feet long. Wild black currants transplanted into the garden this spring show vigorous growth and fruit much larger than usually found when the bushes grow wild...Prolific large red currants received from the east, all lived and all show a very vigorous growth. Six slips of black Naples currants all grew and have sent out shoots three feet in length. Two dozen Haughton gooseberry plants received all grew luxuriantly. White clover sown in the middle of April has made a full growth and flowered this season. The flowers are now going off. Tobacco plant grown without forcing in a hot bed has leaves over a foot long, but is not nearly as good as last year when it was forced at the start. A magnificent array of flowers is included in the attractions of Mr. Knowle's garden.[8]

J. Knowles, who partnered with the better-remembered Thomas Henderson in the early 1880s in two horticultural ventures, a market garden on what is now 98th Street near the city centre and a beekeeping and honey-producing business, exemplified the experimental approach to plant selection promoted by the *Bulletin*.[9] Oliver knew that plant trials played a critical role in the push to expand the repertoire of plants grown in the Edmonton area and he ensured that all successes received coverage in his newspaper. On September 22, 1883, for example, in an article on fruit trees, the *Bulletin* reported

that one or two enterprising residents in our own neighbourhood have succeeded in bringing apple trees safely through the last two or three winters and hope to be gladdened in one or two years more by the sight and taste of rosy-cheeked pippins that they have grown themselves. Mr. Gibb's experience with Russian fruits supplies all that was lacking and inspires us with every confidence that these few little trees are but the precursers of waving orchards the fruits of which already make our mouths water in anticipation.[10]

In 1885, the *Bulletin* began to participate in the quest for gardeners' nirvana by maintaining its own trial garden. It "invested in three different apple trees" that it bought from the Renfrew Fruit and Floral Company of Arnprior, Ontario, a company that entered the Edmonton market in 1884 and advertised persistently in the *Bulletin* throughout 1885. The tiny trees were planted next to the *Bulletin* office where they grew for three years before initial results were published: "According to directions they were buried in the garden during the winter, and in the spring were set out close along a fence which shaded them to some extent from the morning sun until about 10 A.M. while they got the full benefit of the evening sun." Without winter protection or pruning, it was reported, they grew annually until, by 1888, "they were about 5' in height with many branches and a very healthy vigorous growth."[11]

Throughout the 1880s and 1890s, the *Bulletin* continued to promote the importance of plant trials to its readers, a position that received federal government support after 1886 when the Central Experimental Farm was founded in Ottawa under the dynamic leadership of horticulturist Dr. William Saunders. On May 10, 1890, the *Bulletin* expressed indebtedness

> to the Central experimental farm for a number of well-rooted slips of apple trees, currant, gooseberry, raspberry and other fruit-bearing bushes, accompanied by directions for planting. The only condition attached being that good care should be given the slips and progress reported. All were wonderfully well rooted and arrived in excellent condition. Some had already sprouted. They were promptly planted. The farm has sent similar bundles of trees and bushes for planting to a large number of residents of Edmonton.[12]

And, as if to keep the spirit of scientific experimentation alive and well in Edmonton, the *Bulletin* continued to publish reports on the successes and failures in its trial garden throughout the 1890s.

One settler who needed no urging to obtain plant material from the Central Experimental Farm was Donald Ross, who deserves recognition as the first Edmontonian to confront his adopted city's horticultural challenges while fully exploiting its advantages. On July 6, 1889, the *Bulletin* reported that

> [l]ast spring D. Ross received four varieties of strawberry plants, eight of each, from the Central Experimental Farm near Ottawa. Being duly planted all the plants of the Captain Jack and Chester varieties and one each of the other two grew. This season they are bearing fruit abundantly. Some of the roots have four and five bunches of berries with up to 21 berries in each bunch. The berries were ripe on July 1st and measured from two to two and three quarters inches around.[13]

Two years later, Ross "presented the *Bulletin* with a plate of ripe strawberries, averaging three and a half inches in circumference." After reporting on the most successful varieties, the article concluded with the following encouragement to readers: "Considering Mr. Ross's uniform success with strawberries there seems to be no reason why they should not be grown more extensively here."[14]

No wonder, perhaps, that in August 1893 Ross's garden was described in great detail in a *Bulletin* article and held up as an inspiration to "intended settlers and visitors to our town." "If you have not seen it," the article urged its readers, "go and see it now." "East, west or south, we hold, you will meet with few better cared for plots of ground where on a small scale potatoes, cabbages, peas, beans, onions, tomatoes and strawberries are raised to the perfection they are here."[15]

Ross's garden continued to produce prize-winning results throughout the 1890s and its owner, who was also the proprietor of Edmonton's first hotel, drew liberally from it to feed his paying guests.[16] By 1904, when William Paris Reeves made his own leap of faith, arriving in Edmonton from Somerset, England, with a wife, six children, and the expressed intention of becoming a retired hobby gardener, Ross was a crusty pioneer and something of a legend.

Gladys Reeves, fourteen years old in 1904 and the youngest of the Reeves children, told the story of her father's first meeting with Ross.

Reeves was sixty years old when a doctor in England advised him that a sea voyage would be good for his health. According to his daughter, Reeves decided to leave England and chose Edmonton because of its gardening potential, a small indication perhaps of the efficacy of Clifford Sifton's aggressive promotion of immigration during his tenure as Minister of the Interior, 1897–1905.[17] Shortly after their arrival, the Reeves family moved to a house on 101st Street, downhill from what is now the *Edmonton Journal* building. Reeves had barely begun to turn the soil for his first Edmonton garden when he received a visit from Ross. According to an account left by Gladys Reeves, "Mr. Ross" was somewhat less than encouraging to the new arrival, declaring that "[n]othing but taters and cabbage grows here."[18] Was Ross protecting his pre-eminence in the garden or did he simply wish to inject a note of realism into the expectations of a gardener whose experience had been gained in southern England? Whatever the reason, Reeves, far from being dissuaded, went on to successes that rivalled those of his reluctant mentor.

In 1907, Edmonton photographer Ernest Brown, for whom Reeves was then working, took a series of photos of the Reeves garden. They add credibility and substantiality to Reeves's own accounts of the gardens her father grew:

> the things he grew were a matter of daily comment to passers by and during the nine years he lived in the brick house at the foot of the steps below the Journal, the Board of Trade made his garden the show place to visitors contemplating settlement here, to let them see what could be grown, and the wonderful fertility of the soil.[19]

Brown's photos suggest that Reeves began by planting what was known to do well in Edmonton's soil and climate. Mammoth root and vine vegetables dominate these early images. It was not long, however, before Reeves was having considerable success with the flower seeds he ordered every winter from Carters Seeds in England—Canterbury bells, carnation, godetia, petunia, verbena, stocks, sweet peas, and zinnia, to mention a few. In February 1935, when he was ninety-one

W.P. Reeves in his 101st Street garden, Edmonton, 1907. [PAA B.6819]

years old, Reeves sent one pound, two shillings, and sixpence with his sixty-sixth annual seed order to Carters in England, asking for the usual prompt attention to the order as "time will be getting short for spring planting."[20]

A few years before her father's death, which occurred days before his ninety-fifth birthday in October 1939, Reeves wrote a letter to the managing editor of the *Edmonton Journal*, describing her father's contribution to his adopted home. "[M]y father has not been a Public servant, a Politician, or a Public man" she wrote, but "he has perhaps contributed no small amount to the upbuilding of this country, through his hobby—the garden."[21] And, indeed, despite his quiet and unostentatious manner, Reeves had achieved renown for his gardens in a community that, by 1939, had progressed considerably beyond its pioneer beginnings on the gardening front.

On one front, however, Edmonton had not progressed much further by 1939 than it had in September 1883, when Oliver had eagerly anticipated "waving orchards" of "rosy-cheeked pippins," or in 1904, when William Paris Reeves set about planting his first garden.[22] Although the *Journal* noted in 1939 that Reeves "had one of the first apple trees in the district when, in 1907, he was awarded the gold-and-silver trophy at the Edmonton exhibition for a showing of his garden produce," no detailed description of the apples was forthcoming.[23] And, alas, it is in the area of tree fruits that the booster line, so prominent in early writing about Edmonton's gardeners and gardens, falters.

In September 1883, an article in the *Bulletin* notes, almost ruefully, "[w]e have already such an abundance of small fruits that nothing but the ambitious second to none spirit that pervades the North West would justify us in asking for more."[24] Two years later, a *Bulletin* article in March 1885 applauds the agricultural society for "taking up the matter of fruit raising," arguing that the "lack of large fruit—especially the apple, which is the best of all," is perceived as an important drawback for the community, the removing of which would be a "most laudable object."[25]

It is perhaps when we get to apple growing that it is appropriate to mention a second reason why Edmonton's early gardeners set out to prove their city was a gardening Mecca. Proving something to others is always an operative principle for boosters. Satisfying one's own craving for the familiar and the loved, whether it be flower, tree, shrub, fruit, or vegetable, is another powerful motivator for immigrant gardeners. Edmonton's early settlers came from apple-growing places—Ontario, England, France, the United States of America, and Ukraine, this last of which, along with Siberia, was the best hope as a source of seed for apple growing in northwestern Alberta. Apples, above all other edible things, were missed by Edmontonians and no other deficiency in the gardening potential of the locality has received as much attention from governments, organizations, and individuals.

The attempt to grow plants from seeds one has brought from "home" must have helped many immigrants, especially women, adapt to a new environment. That is why Louise, a character in one of Georges Bugnet's novels, planted a garden from seeds she had brought from France:

> [T]his little piece of ground was the only corner of the homestead she could love without reserve. All around her, friendly flowers consoled her lonely eyes. They recalled her homeland.[26]

In his novel *The Forest*, author and plant breeder Bugnet, who came to the Edmonton area in 1905 from Dijon, France, and settled on land northwest of the city near the present-day hamlet of Rich Valley, tells a story that revolves around the humanly experienced conflict between the aesthetics of nature and those supplied through culture. It is the story of Roger and Louise, who leave France to homestead northwest of Edmonton near Lac Majeau. The relationship between man and wife is strained by their different responses to the landscape. Roger is overwhelmed by the beauty of the natural world in which he finds himself, while Louise is intimidated and appalled by it. This difference in response persists. Louise's only relief and escape is to plant a garden with seeds she has brought from France. Her attempt to cultivate the soil and endow it with features that recall her French homeland comforts her at first, but when the plants are all killed by a spring frost, cutting off her supply of seed, Louise begins to hate the place her husband has come to love.

Bugnet lived and worked in the Edmonton area for the rest of his long life; he never returned to the land of his birth. Like his character Roger, Bugnet was thrilled by the natural world he encountered in Alberta. But like his character Louise, he missed the gardens he had left behind in France, and especially some of the flowers that grew in them. Bugnet's admiration for his new surroundings did not inhibit his drive to modify or even change them. As a practising Roman Catholic, Bugnet's explorations in the field of plant genetics were carried out in the conviction that he was helping to make God's wisdom manifest to future generations of western Canadians.[27] Within a decade of his settling in Alberta, this energetic scholar/farmer had become familiar with the native flora, had written away to botanic gardens around the world to obtain non-native seeds that had some hope of surviving the climate, and had begun the experiments that led, eventually, to popular introductions like the 'Thérèse Bugnet' rose.[28]

Edmontonians brought with them a love of roses and the fact that our climate was unsuited to the varieties they especially admired only piqued their interest in growing them. On August 23, 1890, the *Bulletin* reported that "Mrs. Thos. Henderson of Fraser Avenue has a beautiful climbing rose in full bloom. It is grown in the open garden."[29] In August 1911, the display of roses at the Strathcona Horticultural Society's flower show was said to contain standard varieties such as 'American Beauty', as well as moss roses.[30] In the 1920s and 1930s, Edmonton Horticultural Society stalwart H.W. Stiles wrote articles in the *Bulletin* on gardening, articles in which he gave ample coverage to his particular interest in growing roses.[31] From the mid-1930s to the mid-1950s, manager of the Capitol Theatre and rose enthusiast Walter Wilson teamed up with seedsman and rose enthusiast Alfred Pike to make Edmonton the "Rose Capital of Western Canada." Together

'Thérèse Bugnet' roses on 100th Avenue and 116th Street, Edmonton, summer 2013.
[Photo courtesy of the author]

they prepared an instruction booklet on planting and caring for roses. Pike looked after distribution by growing roses that he imported from Scotland before grafting them onto hardy rootstock; he began selling these in vast quantities in 1932. Wilson took care of promotion by holding, every year from 1933 through 1955, an annual rose show in the lobby of the Capitol Theatre.[32] Between 1950 and 2000, the increasing availability of hardy hybrid roses developed by both amateurs and professionals had transformed gardening in Edmonton so that even the least tended of gardens could bloom from June into September. By 2000, hardy shrub roses had become a staple of public, low-maintenance gardens to console the eyes of all Edmontonians.

When George Shewchuk, author of *Roses: A Gardener's Guide for the Plains and Prairies*, 1999, began to grow roses, nostalgia played no part in his motivation. Shewchuk was responding to the challenge that animates every gardener's soul; he wanted to prove to himself that it could be done. As he explained in the summer of 2002, when visited by the author in his southeast Edmonton garden, he had been growing fields of gladioli out of interest but was tired of drying and cleaning the corms in the fall before dusting them with fungicide for storage over the winter. Unaware of those who had travelled the same route before him, he began to experiment with roses in the 1950s, piqued by a convention that held they were unsuited to the Edmonton area. And this is perhaps yet another reason why Edmonton's early gardeners, among whom Shewchuk might be counted, as he was born in 1913, set out to prove that their city was a Mecca for gardeners—to prove, not to others but to themselves, that it was so.

Shewchuk's abiding interest in growing plants, and especially in growing roses, developed so naturally as part of his background that it took him awhile to recognize it as a mission. As a young boy, he worked with his mother in the garden on their farm northeast of Edmonton and helped her with other tasks around the farm. He liked all farm work but had a particular liking for plants and gardening. In addition to horses, cattle, and pigs, the family kept sheep that were sheared for wool. His mother washed, carded, and spun this wool and from it they knitted many of their clothes and blankets. From his mother, young George learned to knit mittens, socks, sweaters, and other articles. His mother and grandmother both grew hemp. The fibre was used to make horse blankets and the seed was pressed for its fragrant oil, as were sunflower, flax, and poppy seeds.

After a seven-year stint during the Great Depression as a schoolteacher, and because he was disqualified for service overseas during the Second World War, Shewchuk enrolled in the University of Alberta's Faculty of Agriculture in 1941 and went on to a career as a district agriculturalist with the Government of Alberta. His special interest was farmstead planning and beautification and, while preparing a booklet on the subject for the Department of Agriculture in the 1940s, he articulated a gardening philosophy that served him for the rest of his life.

> Most individuals have a love for the beautiful in nature. To be surrounded by it, has a most refining and elevating effect. The homes we have

built, the walks and drives we have located, the trees and shrubs we have planted and the beauty we have created about our homes will be ours in a sense they could not be if others had done it for us...Happily, anyone may devote his interest and artistry in perfecting the beauty of some piece of ground. The rewards of our creative effort grow into the most precious source of joy and comfort in later years.[33]

In 2002, at the age of ninety, Shewchuk had 168 rose varieties in his garden, all of which he had planted and tended himself and many of which were very old. In addition to roses, there were many other flowering and leafy plants in the garden, including several varieties of dahlias, some unusual variegated geraniums, columbine, clematis, begonias, poppies, and marigolds. Volunteer marigolds were encouraged where they were wanted. Most of the twelve fuschias he had purchased in Victoria twenty-five years before hung from pots around the edge of the cement patio, propagated each year from cuttings, each one a different combination of colours and each with a name to prove its pedigree. Vegetables were grown in large pots or sprinkled throughout the garden. There were many fruit-bearing trees and bushes such as apple, Nanking cherry, and gooseberry. Next to the back fence, a space between a linden tree and an oak tree sheltered several tall rose bushes, including a 'Thérèse Bugnet'. The linden and the oak, both of which were kept immaculately pruned to keep them clear of the power lines, came from a nursery in Saskatchewan and the oak was grown from an acorn. Every plant had a name and a story, including, of course, the 'Thérèse Bugnet', whose breeder, that other Georges in Edmonton with a passion for roses, was well known to Shewchuk.[34]

Why does anyone plant a garden in Edmonton? To combat the eastern prejudice, perhaps, although in the twenty-first century the same prejudice is more likely to originate from the west coast. To console oneself for what one has left behind, as Edmonton's early Chinese truck gardeners did when they grew cabbages and carrots for their customers alongside bok choy and lo bok for themselves. Or to please oneself by perfecting "the beauty of some piece of ground."

Lois Hole, gardener extraordinaire, entrepreneur, author, and much-loved Lieutenant-Governor of Alberta from 1999 to 2005, who came to prominence more than a century after Edmonton's first gardens were planted, understood all these reasons. Hole shared the optimism of Edmonton's first settlers, always looking forward to the next new hybrid or the next new product and always hoping to be able to do tomorrow what was impossible today. Gardening, for Hole, was a metaphor for life. In the foreword to her son Jim Hole's book, *What Grows Here*, Hole concluded, "In the end, I guess we do. And isn't that why we garden in the first place?" If she had been asked to answer the question "Why grow here?" her answer would likely have been in the same mode. We grow here because we live here.[35]

2

Donald Ross

Edmonton's "Father of Gardening"

The assertion...in the Bulletin that new potatoes were as large as hen's eggs, leaving it to be inferred that there were none larger, has called forth an indignant protest from Mr. D. Ross, who rises to explain that he had at that date potatoes as large as a man's fist and over three-quarters of a pound in weight.
—EDMONTON BULLETIN, August 4, 1883

What do you think of potatoes weighing 4½ pounds?, of onions measuring in circumference 15½ inches—of cabbages and cauliflowers so big you cannot get them into a kettle, some of which will tip the scale at considerably over 30 pounds? Surely this speaks for itself.
—EDMONTON BULLETIN, August 14, 1893

FOR DONALD ROSS, who strove to live up to his deserved reputation as Edmonton's first and pre-eminent gardener, bigger was always better when it came to vegetables. Energetic, enterprising, inquiring,

Donald Ross, 1904.
[CEA EA-10-669.2]

and the embodiment of the booster mentality, Ross was well known and respected during his lifetime for the gardens he planted and for the produce that came from them. On June 29, 1893, by which time Ross had lived in Edmonton for over twenty years, the *Edmonton Bulletin* described him as the town's "pioneer gardener" and "still one of the most, if not the most successful." Ten years later, and still gardening in his inimitable promotional style, Ross was elevated to the status of patriarch when an article in the *Bulletin* admiringly referred to him as the town's "father of gardening."[1]

In August 1872, when Ross arrived to pan for gold in the North Saskatchewan River, Edmonton was in the first stages of a critical transition from fur trade post to permanent settlement. Although the Hudson's Bay Company (HBC) had given up its claim to governance in 1870, the newly constituted Dominion of Canada had yet to establish replacement structures and institutions. The three-thousand-acre land reserve around Fort Edmonton that formed part of the company's agreement with the government had not yet been surveyed, forcing settlers to make educated guesses when they chose a piece of land on which to build. When Ross arrived, the Reverend George McDougall

Donald Ross homestead, 1890. [CEA EA-10-602]

had just chosen a location on the east side of what became the 101st Street eastern boundary of the HBC reserve and was in the process of building Edmonton's first Protestant church there. McDougall's Methodist mission became the nucleus of the Edmonton settlement in its early years, with Ross strategically located on the river flats below it.

Ross was born in Dumfrieshire, Scotland, on June 17, 1840. In 1846, he moved to England with his parents, where his father worked as a gardener. At age thirteen he entered domestic service, taking a position his parents must have judged to have more vocational prospects than gardening. However, before he was seventeen years old, Ross had signed up to work as a cabin boy on a transatlantic steamer and, after only two ocean voyages, he decided to seek his fortune in North America. He worked for three years in the hotel trade in New York City before tales of the California gold rush drew him westwards. From California Ross made his way first to Carson City, Nevada, and eventually to the Cariboo district of British Columbia, before journeying to Edmonton via Peace River and St. Albert. Edmonton might well have been a stop on the road for an itinerant fortune seeker, if, for reasons he never felt it necessary to explain, Ross had not decided to stay. And, if it was gold that drew him to the place, and coal, hostelry, and real estate that sustained him, perhaps it was the garden that held him.[2]

How Ross established himself as a landowner, coal miner, hotel operator, and gardener on land currently known as the community of Rossdale is not absolutely clear. John Patrick Day, relying partly on Ross's own accounts as they appeared in the *Bulletin* or were verbally related to historian Archibald MacRae, has come closest to assembling an account of the sequence of events leading to Ross's transition from itinerant gold miner to established landowner.

While still a child, Ross learned to garden from his father. Presumably it was this early training he relied upon when, in 1874, he struck an agreement with the HBC to operate the company's farm, located on the river flats just outside the HBC reserve. According to Day, "after a year Ross understood that he owned the land outright in return for waiving his operator's fee and [he] opened a hotel on the property."[3] MacRae's first-hand and somewhat credulous account captures some of Ross's characteristic narrative style. In MacRae's version, although Ross had taken a three-year lease on the HBC's farm, the "inspecting chief factor... agreed to cancel his lease" after the first year and to buy back Ross's implements, stock, and crop. "In three weeks," wrote MacRae, "Mr. Ross had all his debts paid and eight hundred dollars cash to his credit with the Hudson Bay Company, besides a part of the crop."[4] Whatever the precise story, Ross was no longer working for the HBC in 1876 when he decided to open Edmonton's first hotel on the former farm site.

The Edmonton Hotel, a community landmark until the spring of 1928 when it burned to the ground, played an important role in the early life of the burgeoning settlement. Under Ross's proprietorship and management, it served the needs of travellers for good food and simple accommodation and offered the kind of congenial ambience that made it a popular choice for local meetings and social events. Assisted by his wife, Olive, whom he married in 1878, Ross undertook some major hotel renovation and expansion projects and presided

Edmonton Hotel and Annex, 1911. [CEA EB-26-265]

over comings and goings at the hotel until the summer of 1892, when, except for occasional events such as the Old Timers' Dinner held there in November 1894, the hotel closed for almost a decade. However, sometime between 1876 and 1892, when the productive capacity of his garden must have greatly exceeded his personal and business needs, Ross decided that he would rather be a gardener with a hotel than a hotelier with a garden. He never changed his mind. Although he reopened the hotel on November 20, 1901, it was with the assistance of a full-time manager. The reopening, which received a brief notice in the *Bulletin* two days later, was celebrated in Ross's characteristic style by a group of locals with band music and a midnight supper.[5]

Only a few of the many references to Ross's garden that appear in the *Bulletin* in the 1880s and early 1890s mention the hotel's table as the ultimate destination for its produce. There is no doubt, however, that the garden added variety and sophistication to the fare Ross was able to offer his guests, and it was a feature of the business he was quick to capitalize on after the *Bulletin* began publication in December 1880. Early in 1882, Ross advertised his "Pioneer House of Entertainment" with the claim that "[p]emmican and dried buffalo meat has long been a stranger at the table, and its place has been taken by substantials more in keeping with the onward march of civilization."6 And, in its column on items of local interest, the *Bulletin* kept readers apprised of the direction in which civilization was heading. "Green peas for dinner at the Edmonton hotel last Sunday, being the earliest date Mr. Ross has had them" the *Bulletin* announced in the summer of 1884, or "D. Ross of the Edmonton Hotel has marrowfat peas from the second blossoming of the vines this season the pods of which measure five inches in length and are well filled."7

During this same period, Ross emerges from the pages of the *Bulletin* as a man who prided himself on producing the first or the biggest or the most shapely of any number of vegetables and small fruits and who thrived on the annual competitions organized by the Edmonton Agricultural Society. In effect, Ross's garden became the ultimate expression of his strengthening faith in the future prospects of his chosen home.

It is possible, by reading between and around the lines in the many references to Ross's garden and its produce in the 1880s and early 1890s, to imagine some of the excitement pioneer gardeners must have felt as they planted their gardens each spring. Where the HBC had repeatedly, and for the most part successfully, planted modest kitchen gardens for the benefit of those who lived at Fort Edmonton and grown field crops of potatoes, turnips, and cabbages alongside grains on its farm, Ross and a few of his neighbours, eager to experiment, planted as great a range of produce as they could obtain seed for.8

Short supply of garden seeds was, however, something of a problem in the early 1880s. In late April 1882, the *Bulletin* announced that "garden seeds are scarce this spring," a scarcity that was somewhat rectified when the same newspaper received a shipment of seed the following week.9 A few months later, in September 1882, the Edmonton Agricultural

Society linked its failure to hold exhibitions in 1880 and 1881 to the seed shortage. Although some of the proceeds of the 1879 show had been used in 1880 to purchase "new varieties of field and garden seeds," these had not arrived in time for the 1881 planting. It was not until the fall of 1882 that the state of farms and gardens in Edmonton warranted holding the exhibition that, as it was afterwards reported in the *Bulletin*, justified the faith of all concerned in their view that Edmonton's gardening future would be a rosy one.[10]

In the 1880s, Ross grew red, yellow, white, and seed onions; potatoes of several sorts; carrots; parsnips; turnips; beets; mangold wurzels; cabbage; cauliflower; peas; radishes; lettuce; cucumbers; citrons; pumpkins; corn; and tomatoes. By 1889, he was successfully growing cultivated strawberries from the Central Experimental Farm in Ottawa and keeping careful track of the productivity and hardiness of each variety. In May of 1891, the cherry tree he planted in his yard in 1889 was cheerily reported as being "in full blossom." Published results of the prizes awarded each fall at the agricultural shows suggest that Ross's hearty garden repertoire was both attuned to the tastes of the public he catered to and a reflection of his own tastes and aesthetic.[11]

Varietal names for the vegetables grown by Ross and other new settlers rarely appear in the earliest newspaper accounts of the period, with potatoes and onions being the occasional exceptions. Potato varieties grown in the 1880s included the popular and wonderfully named 'Beauty of Hebron' and the equally popular 'Early Rose' potato. The renowned American nurseryman Luther Burbank, who came to Edmonton in 1894 and subsequently corresponded with Ross, gained his first big break when one of his 'Early Rose' potatoes sprouted a seed head. This rare event led Burbank in the early 1870s to the successful development of the famous 'Burbank' potato, a variety that appears not to have been grown in the Edmonton area.[12]

Ross was always proud of his potatoes, some of which achieved impressive proportions. In September of 1886, for instance, the *Bulletin* described two of Ross's 'Beauty of Hebrons', a four-pounder of "irregular shape" and a smoother and more shapely specimen that weighed in at three and a half pounds. Onion varieties grown by Ross always included 'Yellow Danvers' and 'Red Weatherfields', both of which are mentioned in the *Bulletin* as early as 1881. However, when the *Bulletin* ran a story on Ross's garden in 1893, varietal preferences were indicated in broad, generic terms. Ross said he liked to grow "large flat dutch" cabbage and a variety referred to as "express." He grew "golden wax, broad windsor and long pod" beans and admitted that the peas for which he was frequently mentioned came from "seed of the original stock he found when he came here and seem specially fitted for the climate."[13]

Varietal names were more carefully tracked after 1886, the year the Central Experimental Farm in Ottawa began to assist Western farmers and gardeners by providing seed for testing purposes and by setting up its own breeding program. In 1894, for example, results published in the *Bulletin* for the fall exhibition held by the South Edmonton Agricultural Society named seven potato varieties. In addition to 'Early Rose' and 'Beauty of Hebron', the newspaper account referred to 'Morning Star', 'Ladies Finger', 'Burpees Extra', 'Rural New Yorker', and 'Beauty of Onondaga' potatoes.[14] A week later, the winner of the first-prize potato collection in the Edmonton show assembled a display of nine varieties, including 'Beauty of Hebron', 'Early Rose', 'Stray Beauty', 'White Star', 'Snowflake', 'Burpees Perfection', 'Northwest Beauty', 'Early Ohio', and 'Blush'.[15] These annual fall exhibitions provided local gardeners and farmers with ongoing opportunities to learn from their own and their neighbours' successes and failures.

Named varieties of cultivated strawberries may have been introduced to the Edmonton area as early as 1884, when G.A. Blake reportedly received strawberry plants from Scotland. A year later, the *Bulletin* noted the success of this experiment and, at the same time, carried advertisements from the Ontario-based Renfrew Fruit and Floral Company, which was selling strawberry plants to buyers in the Edmonton area.[16] It was not until 1888, however, when Ross received four named varieties of cultivated strawberries from the Central Experimental Farm in Ottawa, that the distinction between wild and cultivated varieties began to receive ongoing newspaper coverage. In

1889, the *Bulletin* reported that while all four test varieties had survived in Ross's garden and were thriving in their second season, the hardiest plants were of the 'Captain Jack' and 'Chester' varieties.[17] A year later, Ross took a plateful of berries from these plants to the *Bulletin* office. Readers of the account that appeared in the following day's newspaper would certainly have understood that this size could not be matched by berries from wild plants.[18] By June of 1891, Ross's best-performing varieties seemed to be 'Captain Jack' and 'May Queen', and the *Bulletin* proprietor drew from Ross's success in this venture the lesson that cultivated varieties should be grown more extensively.[19] No surprise, then, that in 1893, the year after he closed the Edmonton Hotel to devote himself more fully to other enterprises, including gardening, Ross was quoted in the *Bulletin* as having decided to double the amount of garden space devoted to his strawberry patch.[20]

Ross's apparent indifference to the wild strawberry, which was prolific in the Edmonton area at the time, as well as being hardy, disease-resistant, and tasty, is worthy of note more than a century later when the terms "wild" and "native" often add value to produce. Beginning in the summer of 1882, the *Bulletin* rarely failed to signal the availability of wild strawberries on the market and even to highlight their abundance in the many articles it ran to promote settlement. Nevertheless, new settlers were cultivators before they were gatherers, and there was a noticeable predisposition against harvesting from nature's garden. "Strawberries are very plentiful," announced the *Bulletin* in July 1882, "but the whites are too busy and the Indians too lazy to pick many of them."[21] A year later, an article on fruit cultivation pointed out that while "blueberries and cranberries, strawberries and raspberries, gooseberries and currants, are to be had for the picking…a rich reward not only in convenience of picking but in size and flavour of fruit is in store for the man who begins to cultivate them."[22] Wild fruits, including the strawberry, "can be procured at the mere cost of transplanting," the *Bulletin* noted in an article in 1888, but it was a recourse advocated only for the farmer with neither the time nor the money to experiment with cultivated varieties.[23]

And so, well over a century after domestic strawberries began to replace wild ones on the tables of Edmontonians, one must seek out the descriptive accounts that appeared with relative frequency in the *Bulletin* to be reminded of what preceded our cultivated gardens. Consider, for instance, that in 1893, when Andrew Osler visited the Edmonton area from his home in Scotland, he was driving north of the Sturgeon River when he noticed many "Indians" and "half-breeds" picking wild fruit to sell. Osler was stunned by the abundance he encountered and he singled out strawberries as being especially prolific. The fruit was so abundant in some places, he wrote, "that in walking along a person's boots are painted crimson, and his footsteps have the appearance of a trail of blood."[24] Nevertheless, even as the *Bulletin* was announcing in the spring of 1894 that "the whole of the open prairies is one vast strawberry patch," its proprietor, perhaps inspired by the experiments of Ross, was happily growing 'Captain Jack' strawberries in his garden.[25]

Ross's involvement with the Edmonton Agricultural Society, as both organizer and exhibitor, illustrates the belief he shared with Oliver that support for the society and its endeavours was the best way to promote Edmonton's agricultural and horticultural potential. When, with Ross's assistance, the Edmonton Agricultural Society reorganized and decided to hold a fall show on October 19, 1882, Ross was ready, winning first prizes for his beets, parsnips, mangolds, red onions, and tomatoes, and second prizes for both his 'Early Rose' potatoes and mixed pickles. In 1883, the Roman Catholic Mission and the Hudson's Bay Company were both big winners at the fall show, but Ross managed to capture firsts for the best six carrots and the best two cabbages and seconds for his citrons and his 'Beauty of Hebron' potatoes. One cannot help but think that Ross's refusal to accept nomination as president of the Edmonton Agricultural Society in 1884, although he agreed to stay on as a director, was because he preferred competing to organizing. In 1886, he won eleven prizes for his vegetables, including a third for cucumbers and a special award for his citrons and, in 1889, he was again a big winner. By 1890, however, there is evidence that Ross's competitive interests had found a new focus.[26]

In retrospect, the two dollars won by Ross in October 1890 for the best "Collection of Vegetables" in the fall exhibition pointed to his increasing interest in the challenge of creating eye-catching displays.[27] He must have been disappointed when, in 1894, Count de Cazes from the Stony Plain Indian Agency surpassed Ross's entry. The count took first prize with an exotic display that included eleven varieties of cabbage; green and Egyptian kale, the curly leaves of which "looked like tropical trees more than vegetables"; tomatoes; melons; citrons; carrots; cauliflower; celery; corn; cress; chives; salsify; spinach; sage; sweet marjoram; tobacco; hemp; mushrooms; chervil; lettuce; cucumbers; asparagus; horse radish; "and all the other more common varieties of garden stuff grown in the district."[28] By contrast, Ross's second-place entry was noteworthy only for its inclusion of "large English beans on the stalk." Perhaps Ross consoled himself in 1894 for his second-place showing on the vegetable front with the first prizes he won for the home-cured bacon and hams he had just begun to pack on a commercial scale. Pork packing was a business in which the *Bulletin* quickly pronounced Ross a "professor," although it turned out to be one that did not hold his attention as did gardening.[29]

From 1895 onwards, Ross outdid himself as a display artist at any fall exhibition he chose to enter. In the Edmonton exhibition that year, he won thirteen prizes, including one for the best collection of vegetables.[30] A few days later, he made an excellent showing at the South Edmonton Exhibition, where his vegetable display was accorded a feature location. "Inside the building," the *Bulletin* reported, "the display of vegetables by D. Ross was the first thing to meet the eye of the visitor and was a credit to the exhibitor and a show in itself."[31]

In September 1900, Ross was one of two men put in charge of the South Edmonton Agricultural Society's display at the Inter-Western Pacific Exhibition held in Calgary. "Mr. Ross had two exhibits, one in conjunction with the Strathcona Agricultural society and the other for himself," reported the *Bulletin*. "They included: Eighteen pumpkins; eighteen vegetable marrows; twenty cabbages of each of five varieties; nine kinds of onions; 3 kinds of celery, 18 plants each; beets; turnips;

2 kinds of cucumbers; also a bouquet of flowers three feet high and of such dimensions that it was scarcely possible to squeeze it into a flour barrel."[32] By all accounts, the elements of these displays had been poorly arranged for the first day of the show as a direct result of Ross's delayed arrival, provoking the *Bulletin* to note that visitors to the show that day had been deprived of the benefits of Ross's "artistic hand."

In September 1902, Ross again represented the South Edmonton Agricultural Society, this time in Winnipeg at the Western Horticultural Society's exhibition held there in August. The exhibit of roots and vegetables he took with him came entirely from his own garden, the *Bulletin* account boasted, and it was enough to garner a second prize. Praiseworthy though this was, the *Bulletin* felt compelled to account for his failure to obtain a first:

> *Mr. Ross was at a great disadvantage in that his exhibit was the product of only one garden, while his competitors have no doubt been able to secure exhibits from a number of gardens. Besides his exhibit was taken from the garden six or seven days before the exhibition and lost several days growth and deteriorated by keeping as compared with exhibits from Manitoba points.*[33]

Despite this small disappointment, Ross returned to Edmonton triumphant, having won a first for his conical cabbage and received special notice for entering the largest turnip in the show, an eleven-and-a-half-pound specimen.

Ross's confident showing at the Winnipeg Horticultural Society's exhibition in 1902 occurred when his local reputation as Edmonton's pre-eminent gardener was at its highest. "His was the first garden to produce early vegetables for sale," the *Bulletin* reported in January 1903, "and to his greenhouses is chiefly due the early blooming of flower beds in the town, as well as the plentiful array of window plants to be seen and the cut flowers which appear at all seasons suitable for the occasion." In the decade between 1892 and 1902, Ross managed to turn a prolific hobby garden into a market gardening enterprise that was, as described in the *Bulletin*, both "a success for himself and a credit to the district."³⁴ Although it would not be long before Ross's business success was equalled and even surpassed by newer arrivals such as the better-remembered Walter Ramsay, it was Ross who paved the way for his many successors.

The market for Ross's produce extended well beyond the boundaries of his own town. As did one or two other Edmonton-area market gardeners, he frequently shipped loads of produce to Calgary by train. In August 1891, he was reported in the newspaper as having "shipped out a quantity of vegetables on Wednesday's train to J. Brown supply offices on the C&E now south of Calgary."³⁵ By 1894, when his market garden was in full production, Ross regularly shipped vegetables to Calgary at a freight rate of sixty-eight cents per hundred pounds. In July of that year, for instance, the *Bulletin* reported that in one week he had made two such shipments including "cabbages, potatoes, peas, turnips, beets, radishes, etc."³⁶

The importance of the Calgary market to his business notwithstanding, Ross's major influence as a gardener was felt in Edmonton, where he sold garden produce, bedding plants, and even cut flowers from "Edmonton Under the Hill"; delivered "[e]arliest fresh vegetables in season" to his customers; and sold indirectly through some of the town's first retailers.³⁷ "D. Ross has new potatoes of good size for sale at W.G. Ibbotson's," the *Bulletin* reported in early July 1893, or "D. Ross has fresh radishes for sale in his market garden."³⁸ However, by 1894, inspired by the experiments of fellow hotelier Frank Marriaggi, Ross had decided to extend the Edmonton growing season and provide conditions for growing more exotic produce by constructing a greenhouse.

Marriaggi,³⁹ manager of the Alberta Hotel on Jasper Avenue in the early 1890s, and a man whose gardening enthusiasms appear to have been directly linked to his interest in fine dining, liked to claim that the table he offered guests was "unsurpassed" in Edmonton.⁴⁰ In the spring and summer of 1893, the *Bulletin* reported first that Marriaggi had "a green house in running order which is a credit to his taste and enterprise," and then that he had "laid out a most tasty garden in what was formerly the back yard of the hotel."[41] In August 1894, diners at the Alberta Hotel were treated to muskmelons "of the Noire des Carmes variety" grown by Marriaggi outdoors, while more tender varieties were apparently flourishing "under glass."[42] And, always thinking towards the value of local produce for his hotel table, Marriaggi decided for the fall fair in October 1894 to donate "a $5 prize...for the best brewed beer, ale, lager and porter, one of $3 for the best home grown tobacco and another of $2 for the best native wine."[43] In the summer of 1895, the *Bulletin* reported that due to a breakup in the partnership between Marriaggi and a Mr. De Roux, the former would be leaving Edmonton to take over a business in Fort Saskatchewan, a displacement that effectively removed the hotelier from the local gardening scene.[44] Just a few months after Marriaggi's departure, however, Ross, who must have taken great interest in this daring experiment, built a greenhouse of his own "in which to produce early vegetables for market and to ripen delicate products."[45]

Marriaggi's greenhouse may have been the first to be constructed in Edmonton; it was certainly the first to warrant mention in the *Bulletin*. But Ross, who recognized potential when he saw it, seized immediately on the concept and applied it with his characteristic business acumen. In 1900, five years after he had constructed his first greenhouse, he built a matching one. Thanks to an article published in the *Bulletin* in 1903, we know that Ross's greenhouses were a hundred feet long and twelve feet wide, that they lay alongside one another, and that they were heated by hot water supplied by pipes laid the length of both buildings. The water was heated with the assistance of a forty-horsepower boiler that Ross had salvaged from the government sawmill used to cut lumber for the police barracks at Swan River in 1875. The water came from a well on Ross's property "which rises to a higher level than the greenhouses." Well water was directed to barrels standing inside the greenhouses before being heated and distributed.[46]

By 1895, influenced perhaps by his expansion into greenhouse growing or perhaps by a visit that had been paid to Edmonton in 1894 by American nurseryman and plant breeder Luther Burbank, Ross had begun to expand his business from its original focus on vegetables to include bedding plants, cut flowers, and decorative house plants. All through the spring of 1895 and 1896 he advertised "Garden Plants" and "Plants for the Flower Beds" for sale in the newspaper. In January 1903, we know that Ross had dedicated one of his greenhouses to fresh lettuce, while in the other he produced an impressive array of houseplants: "The cyclamen are just beginning to flower," the *Bulletin* reported, and "[c]arnations, roses and fuschias are in bloom for cutting."[47]

The likelihood that Ross's garden repertoire was influenced by his association with Burbank was advanced by Day, who suggested that Ross wrote to Burbank for plants, thereby attracting the "celebrated American botanist" to pay him "an extended and unpublicized visit in 1894" and to send plants for experiment.[48] That Burbank visited Edmonton is certain, for the *Bulletin* announced his arrival in mid-July 1894, describing him as a man who "makes a business of originating new varieties of fruits and flowers by hybridization and other means."[49] In July of the following year, the *Bulletin* reported that Ross had "collected 1000 bulbs of the orange lily to be shipped to Luther Burbank of Santa Rosa, California, for growth on his experimental farm there."[50] If the lilies survived, which is questionable as they are notoriously fragile and difficult to transplant, there is no evidence that they ever made it into Burbank's list of "rare creations." Whatever the nature and extent of Ross's collaboration with Burbank, it was a sign that Edmonton's gardeners were rapidly moving to a new level of horticultural enterprise.[51] By 1900, the year Ross built his second greenhouse, the viability of kitchen gardens could be taken for granted. Market gardeners were taking on the job of large-scale vegetable production and the local shops supplied an impressive array of locally produced and imported foodstuffs. Urban homes required a more decorative approach to gardening and the new urban class of settler came with aesthetic and culinary tastes that went beyond the standard repertoire. The search for the exotic was on.

The Edmonton Ross left behind when he died in December 1915 was a very different place, horticulturally speaking, from the

Vegetable display from Donald Ross's garden, Edmonton, 1902. [PAA B.9021]

Edmonton he encountered when he arrived forty-three years earlier in search of gold. A city of about seventy thousand inhabitants, which spanned both sides of the North Saskatchewan, had sprung up and was in the process of reshaping the landscape. Market gardeners aplenty drove their rigs in to the downtown market square on Saturday mornings to sell their produce to city dwellers. Ramsay, who came to the Edmonton area as a teacher in 1899 but who left teaching in 1905 to indulge his abiding love of horticulture, had been operating Ramsay Greenhouses for nine years on Victoria Avenue and 111th Street, and was well on his way to becoming Edmonton's most well-known and trusted florist. The Edmonton Horticultural Society, which was founded in 1909 with Ramsay as its first president, had added more competitive opportunities to Edmonton's gardening scene with its annual bench shows and garden competitions and was pursuing its mission of city beautification. Alfred Pike had just launched A. Pike & Co., a seed company that grew to become a respected and well-known brand in the West. A Faculty of Agriculture had just been created at the University of Alberta, with George Harcourt, a gardening activist and Alberta's first Deputy Minister of Agriculture, as its first professor of horticulture. The City of Edmonton had created parks, set up its own tree nursery, and was adding boulevards and trees as fast as its budget allowed. Ross's earlier prominence as Edmonton's pre-eminent gardener had given way to a younger generation with new visions of horticultural nirvana.

Still, Ernest Brown's 1902 photograph of vegetables from Ross's garden, which depicts the gardener himself on the right-hand side of the display, evokes a kind of nostalgia for that unrecoverable period in Edmonton's horticultural history. The vegetables, which tower over both the gardener and an almost obscured gentleman on the left-hand side of the photo, do not look as delectable as they do impressive. Most are as substantially sized as is the browned and weathered gardener, who is standing on the right-hand side of the display wearing a small cap and resting his left hand casually on his right hip. In the background, which is almost certainly the Rossdale flats, the landscape has not yet been touched by buildings or by the introduced species of trees and bushes that were already on their way to transforming the city's image.

3 Nature's Garden Transformed

> No prettier scene can be imagined than these prairies covered with a flow of richest blossom cultivated by the hand of nature.
> —*EDMONTON BULLETIN*, February 22, 1894

> Treating culture as something separate from nature has caused historians to lose sight of the fact that modern society continues to be shaped by its natural surroundings.
> —CLINTON EVANS, *The War on Weeds in the Prairie West*

> I am not sure whether it is a cause for laughter or tears that gardens may indeed have to take over from the replication and replacement of the once rich, natural habitats of the earth, but clearly we must cultivate what we once took for granted.
> —JANE BROWN, *The Pursuit of Paradise*

JANE BROWN ends her social history of gardens and gardening, *The Pursuit of Paradise*, wondering "whether it is a cause for laughter or tears that gardens may...have to take over from the replication and replacement of the...natural habitats of the earth." But she sees no alternative: "clearly we must cultivate what we once took for granted." Worried by contemporary gardening trends that emphasize design and sensational plantings above all else, she reminds readers that all gardens "are made of this earth, and it is the only Earth we have."[1] Clinton Evans, who argues that the proliferation of chemical farm products after the Second World War led to the sacrifice by western Canadian farmers of long-developed principles of land stewardship, would agree with Brown that to think of a garden as a form of exterior decorating is to miss the point. "Treating culture as something separate from nature," Evans says, "has caused historians to lose sight of the fact that modern society continues to be shaped by its natural surroundings."[2] Observations such as these provoke questions about the efforts made by Edmontonians over time to promote their city as a place where crops and gardens of all sorts flourish. How did Edmonton's early settlers and visitors react to the natural world they encountered? With what intentions have they sought to modify or transform it? Where is nature in Edmonton gardens today?

"Prior to settlement," wrote George Turner in an article published in 1949 in *The Canadian Field-Naturalist*, "the Edmonton area was largely wooded with aspen...and balsam poplar mixed with willow species." Turner went on to describe the many spaces of open, grassy land that existed between poplar bluffs and noted that both white birch and white spruce grew along the streams and lakes. Turner's article, an inventory of the many plants he had collected within a fifty-mile radius of Edmonton, drew also upon the collection of Dr. Ezra H. Moss, the University of Alberta professor of botany who went on to write the authoritative reference work *Flora of Alberta*, first published in 1959.[3] Turner's, and then Moss's, groundwork eventually led to reference works such as *Trees and Shrubs of Alberta: A Habitat Field Guide*, by Kathleen Wilkinson, and France Royer and Richard Dickinson's authoritative *Plants of Alberta*.[4] Wilkinson's detailed, illustrated entries on native trees and shrubs, which include information on how plants were used by Aboriginal peoples, help lay readers learn what grew here before European settlement. The Aspen Parkland, one of the six distinct growing areas in Alberta and the one in which Edmonton is situated, is described as consisting of "groves of aspen with occasional balsam poplar, white spruce and paper birch. Shrubs such as saskatoon, buckbrush, roses, wolf willow, choke cherry and beaked hazelnut are common, while wild sarsaparilla, asters, goldenrod, wild vetch, Solomon's seal, pea vine, violets, fairy bells, tall mertensia and cow parsnip are frequent in the herbaceous understory."[5]

Soil quality, judged to be consistently high by early settlers, varied from one part of the Edmonton area to another. Mapping did not begin until 1920, after F.A. Wyatt joined the Soils Department of the Faculty of Agriculture at the University of Alberta, and it was many years later before information was readily available about soil composition in specific areas.[6] Nevertheless, Edmonton's soils produced impressive gardens and crops for the early settlers and the *Edmonton Bulletin* was a convenient vehicle for conveying this fact to prospective settlers. "The soil is generally a vegetable mould averaging two feet deep, resting on a light yellow clay subsoil, and can be worked with the greatest ease," boasted the *Bulletin* in July 1882. "It never becomes hard."[7]

Native berry fruits, which grew abundantly in the Edmonton area, were considered a strong point in the suit to attract new settlers, although such fruits were often presented as mere indications of what the land could be made to produce by an enterprising farmer or gardener. "Strawberries, raspberries, high and low bush cranberries, saskatoon berries, blue berries, goose berries, black and red cherries, black currants, and many other kinds of small fruits abound," reported the *Bulletin* in 1882, before hastily going on to say that while "[h]ops are not native, [they] grow luxuriantly wherever planted."[8]

The perceived harshness of Edmonton's climate was downplayed. In April 1884, the *Bulletin* reported that poplars "near the houses" and along the edge of the riverbank were budding, "while in other localities the buds have hardly begun to swell yet. This proves," was the prescient conclusion, "that thick settlement will make a change in the climate."

Travellers to Edmonton, at least those whose reports appeared in the *Bulletin*, rarely failed to be impressed by the natural world they encountered. Andrew Osler visited the Edmonton district in 1893 from Scotland, for instance, and returned with a most favourable account. Writing for the *Dundee Courier* in January 1894, Osler waxed lyrical on the subject of a drive he had taken north from Edmonton to St. Albert the previous summer. In large open spaces between the wooded groves he was surprised to find the ground "carpeted with wild flowers of every hue." Wild fruits were so abundant that when his party stopped to eat a picnic lunch it was able to put together a "sumptuous and delicious dessert of wild fruits picked where we squatted." Osler's evident appreciation for the potential of the area as future farmland did nothing to diminish his appreciation of the unmodified landscape. He ended a detailed description of his day's journey by declaring that "[n]o prettier scene can be imagined than these prairies covered with a flow of richest blossom cultivated by the hand of nature."[9]

Osler's visit to the Edmonton area came more than twenty years after Edmonton's first settlers had begun to plant the fields and grow the gardens that were to alter the local landscape. And these settlers, all of whom were as impressed as the visitor by the evidence of soil fertility, were nevertheless impatient to connect the dots and colour in the squares of what was to be Edmonton's gardening future. It is often not clear what they saw in nature or whether they viewed it as complementary or hostile to their gardening aspirations. They were certainly not willing to live with it on the same terms Aboriginal peoples before them had done.

Prior to the arrival of the fur traders at the end of the eighteenth century, and settlers just a century later, Aboriginals had occupied for centuries the territory we now know as Edmonton and its surrounding area.[10] Evidence suggests that these Aboriginal groups were hunters and gatherers, not farmers, and that they were highly mobile. Meat formed the bulk of their diet, with plants playing a supplementary but important role; berries, roots, barks, and leaves were used for food, medicines, dyes, and a variety of other purposes. "To survive," wrote Anne Anderson, writing about the Cree side of her ancestry,

> meant roaming from place to place, on the plains, woods and near lakes...Meat was used fresh, dried, pounded for pemmican. Animal fat, berries, edible roots, wild teas such as mint, labrador tea (muskeg tea) and yarrow. [sic] Saskatoons and chokecherry were the main berries, being dried or crushed.[11]

Traditional knowledge included knowing where to find specific plants, understanding the variety of uses to which they could be put, and being able to transform them into whatever was required. It was important, therefore, that the sources of such necessaries be protected by ensuring that they were not overharvested or otherwise destroyed.

W.R. Leslie, superintendent of the Dominion Experimental Farm at Morden, Manitoba, from 1921 until 1956, referred to the manner in which Aboriginal peoples understood the plant world and used it to their advantage without significantly disturbing it as gathering "from Nature's garden." "The prairie Indian enjoyed the beauty of flowers but did not mar the plants by picking them," wrote Leslie, before going on to explain the various practical uses to which the "seeds and fruiting bodies" of plants were put by Aboriginal groups.[12]

In the late 1970s, the University of Alberta's Devonian Botanic Garden began research leading to the eventual opening of a Native Peoples Garden in July 1982, a job that proved to be more difficult than anticipated according to then director Patrick Seymour. Seymour and his staff worked with Anderson, a Cree-Métis who lived in Edmonton and whose life work was invested in restoring the language and traditions to her people.[13] Anderson transcribed material as told to her by the Cree Luke Chalifoux and the result, *Some Native Herbal Remedies*, describes forty plants used regularly by the Cree, primarily in medicinal applications. "The Indian," the introduction explains, "lives very close to nature and knows what is offered to him."[14]

For almost a century prior to the 1870 settlement between the Hudson's Bay Company (HBC) and the Canadian government, the economy and development of the West revolved around the fur trade. In the early years of the trade, there is apparently little evidence that trading post employees systematically grew gardens. This changed after 1810, when a new HBC policy dictated that posts supply as much of their own food as possible. Beginning in 1814, according to Ted Binnema and Gerhard J. Ens, "the journals of Edmonton House are replete with references to farming and agriculture."[15] After 1821, the year the North West and the Hudson's Bay companies amalgamated, daily journals kept at Edmonton House testify to the successful growing of barley, potatoes, turnips, cabbages, and, when the weather conditions were favourable, wheat. These items, planted as field crops, supplied those who lived at the post and were sometimes sent to provision other posts. Company employees, assisted primarily by the women and perhaps the older children, planted, tended, and harvested the crops, or gardens as they were sometimes called.[16] On October 14, 1826, for example, the daily journal kept at Edmonton House records that "with the addition of the Indian women we secured to day about 1000 cabbages and put them into one of the old buildings."[17]

Gardening activities generally began at Edmonton House as early as April when cabbages were started in hotbeds "made for that purpose."[18] Kegs of potatoes from the previous year's harvest were cut into seed and planted in April and May. Turnips were sown from seed. Barley and wheat were sown as early as April, but while barley was a reliable crop, wheat was not. In September 1827, Chief Factor Rowand complained: "five of our men [were] employed cutting the wheat to obtain straw more than the grain which is worth nothing."[19] In addition to the various field crops, references to a "kitchen garden" appear frequently in the journals, suggesting that vegetables were planted in a protected garden near or within the walls. For several years in the 1820s and early 1830s, this garden seems to have been presided over by one of the older employees, referred to only as Valle, who was assisted in his labours and in his weeding of the potato crops by the women and occasionally children of the fort. On April 29, 1833, for instance, a journal entry refers to him as "Old Valle" and notes that he "keeps busy at his little vegetable Garden."[20]

The few pictorial representations extant, which date from well after 1832 when Fort Edmonton was relocated from the flats to a location just under the current Alberta Legislature, suggest the disposition of the fields in relation to the buildings. An ink and watercolour sketch of Fort Edmonton by Mervin Vavasour, 1846, clearly shows the existence of "fields."[21] Ernest Brown's photo of an unidentified painting of Fort Edmonton, dated 1867, shows fenced fields and outbuildings to the east of the fort and suggests a cultivated area on the northwest slope above it.[22] These fenced and cultivated areas outside the walls of the fort, planted with grain and root vegetables on a scale that would be similar to a contemporary market garden, were regarded by both traders and their Aboriginal trading partners as sources of supply. On matters of husbandry, however, traders and the tribes they traded with rarely saw eye to eye.

The groups of Aboriginals who routinely camped for periods of time on the flats outside the walls of Fort Edmonton to trade were considered by HBC employees to be more hazardous than the weather to the well-being of crops. A report filed with the HBC in 1862 from Fort Edmonton notes the "cultivation around the Fort" and the "fair crops" of barley, potatoes, and wheat, but it concludes with a reminder that "[f]arming at this place is conducted under many difficulties, from the number of Indians who resort here, occasionally in large bands and often not too well disposed, large quantities of potatoes are stolen, Fences broken down [and] burnt and their Horses let into the fields to feed on Barley and Wheat."[23]

Ten years later, when George Grant wrote about the 1872 Sandford Fleming expedition, he noted that Aboriginals camping near Fort Edmonton were the primary hindrance to what would otherwise have been a successful large-scale gardening enterprise:

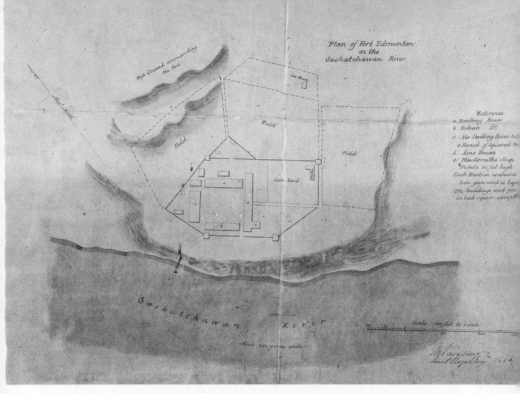

Plan of Fort Edmonton on the [North] Saskatchewan River. M. Vavasour, Lieutenant Royal Engineer, 1846. [HBCA G.1/192]

The usual difficulties from the Indians camping near a Fort have been experienced. A band of strange Indians come along, and, without the slightest idea that they are doing anything objectionable, use the fences for tent poles or fuel; and their horses then getting into the fields destroy much of the crop. But in spite of these and other hindrances, a thousand bushels of wheat are usually stored from a sowing of a hundred; and last year two hundred and fifty kegs of potatoes (eight gallon kegs used instead of bushels) were planted, and about five thousand were dug. The same land has been used for the farm for thirty years, without any manure worth speaking of being put on it.[24]

Initially, the fur trade economy brought with it no imperative for Europeans and Aboriginals to understand and adopt one another's ways of living with the natural world, no need to choose one way or the other.

Each group benefitted from the other's approach, with traders adopting Aboriginal hunting practices and methods of preserving dried meat by pounding it with dried berries to make pemmican, while Aboriginal peoples came to appreciate root vegetables, potatoes, and grains and to use them in their diets whenever they were in trading proximity of a fort. Settlement, with its inevitable assertion of land rights, put nature's garden in jeopardy.

When Georges Bugnet arrived in Alberta, settling in early spring, 1906, on a homestead northwest of Edmonton between Lac Ste. Anne and Lac La Nonne, he came into close contact with Aboriginal groups, including Métis, who lived in the area. Less than two decades later, when he published his second novel, *Nipsya*, he demonstrated some of what he had learned from his neighbours. Bugnet's novel, set near the author's homestead around the time of the 1885 Northwest Rebellion, is named after its central character.[25]

Nipsya, whose name is explained as signifying "the willows" in Cree, is a young Métis girl who has been raised by her Cree grandmother. Her intelligence and sensitivity are largely depicted through the knowing and respectful way she goes directly to nature to supply her needs. In the late summer, for instance, she helps her grandmother gather their "annual supply of fruit." They know exactly what they are looking for and where to find it:

> *In places close to the lake the prickly gooseberry bushes were laden with green or purple berries. In the damp sloughs and on the thick moss of the muskegs, red with* atocas *that were yet too hard and bitter, they picked black currants and sometimes white, purple, and red currants as well, that had not too bad a flavour. Small red cherries and bunches of black choke-cherries glistened on the sunny slopes, blueberries were plentiful in the sandy soil where pines flourished, and beneath the arches of the vast forest,* pembinas *and* graines d'orignal *displayed their umbels of scarlet berries so much enjoyed by partridges and all the other winged folk of the woods.*[26]

Nipsya honours the Cree view that food, medicines, and other necessaries can be gathered directly from nature, and it is through her eyes that the narrative voice in this novel presents nature and landscape to the reader most vividly. At the same time, however, the author shows the reader that he both recognizes and welcomes the culturally transformative effects that agriculture and horticulture are bound to have on Aboriginals living in the Edmonton area. Bugnet attaches great significance to how his characters understand, and live with, both nature and the garden.

Nipsya's three suitors, one of whom is a Cree warrior, one a white employee of the HBC, and one a Métis, all represent distinctive cultural stances towards nature and these stances influence the success, or lack thereof, with which they pursue their suits. For example, when Monsieur Alec, the HBC factor, attempts to make conversation with Nipsya by describing the landscape before them as "sad" and "lovely," the girl he is trying to impress is thrown back on her own very different reactions to the scene. For her, "sadness" is not an attribute of landscape and "loveliness" lacks specificity. She looks in the same direction as Monsieur Alec, but what she sees is a detailed and endlessly interesting picture that both delights and informs her. It is fall, and the yellow birch leaves stand out against the "dark patches of spruce and pine." Near where they were sitting, "the grass was turning yellow and above it rose the stems of fireweed, whose pink seedpods had opened to release their masses of long, silken, silver threads. Large blue, star-shaped flowers, like wild chicory, on stalks of bronzed crimson, mingled with tiny pink daisies, white-flowered 'goose-grass,' and purple thistles, while close to the low-growing *symphorine* bushes with their numerous little snowballs, were the scarlet hips of the wild rose."[27] Monsieur Alec's insensitivity to the natural world disqualifies him as a suitor and their tentative mutual attraction does not develop.

It is through Nipsya's relationship with her cousin and eventual husband, Vital Lajeunesse, that the author conveys his assumption that once the precepts of agriculture, horticulture, and Christianity have been fully accepted by the Métis population, the garden will spontaneously relocate from nature to the homestead. Their relationship begins when Nipsya and her grandmother journey to spend time with the Lajeunesse family at their homestead near the fictional mission at Rivière aux Reflets Rouges. What first appears curious to Nipsya, a two-acre garden planted with barley and vegetables and surrounded by a zig-zag fence constructed of poplar poles, gains in aesthetic appeal only when she fully understands its purpose and has spent some time in it. First, she is instructed in its contents by Vital, who explains that the seeds came from France and were supplied by an Oblate Father from the St. Albert Mission. Later, after assisting with the weeding, she looks at the cultivated garden with new eyes, noticing that her work "gave the garden a neat appearance and a pattern more striking than the subtle harmony of wild vegetation."[28] Nipsya and Vital share a respect for nature's garden, but they also accept and adapt to cultural forces introduced by white settlement.

Bugnet raises questions about the garden as a site of cultural connection and cultural conflict during the fur trade and early settlement periods of Edmonton's history. Because despite the fictional ease with which Nipsya was converted from gatherer to cultivator, evidence suggests that throughout the fur trade and early settlement period the garden did not transpose itself neatly from natural to domestic settings. As the settlement community relentlessly pursued its vision of the domesticated garden, "nature's garden" began to disappear as a cultural concept.

Hunting and gathering, which had formerly supplied many of the fur traders' needs, played a diminishing role in the food supply of the settlement community, replaced by agriculture, horticulture, and imported goods. Wild berry crops, often reported in the newspaper to be heavy, tended to be all but ignored by white settlers unless they were harvested by Aboriginal women and sold in town. In mid-July 1882, for instance, although "[n]ew potatoes the size of a prairie chicken's egg" were reported as being available in Edmonton, the wild strawberries that were reported to be plentiful at the same time went largely unpicked. In 1886, "[t]ons of Saskatoon berries" were said to have "gone to waste this season right here in the settlement for lack of available labor to pick them." At the same time, the *Bulletin* noted, "[i]n the coming winter there will probably be nearly as many tons of canned fruits brought in from Ontario and California."[29]

The tendency to disparage or to rate as inferior any of the fruits or flowers that grew naturally in the Edmonton area and to promote produce grown from imported seeds and root stock was reinforced by the arrival of seed companies in Edmonton and by the opening of the Dominion Experimental Farm in Ottawa in 1886. In April 1882, seeds were reported to be scarce, a scarcity that was ameliorated by the arrival of "[g]arden Seeds of all kinds" at Frank Oliver's.[30] Oliver continued to receive seeds by mail in the spring for the next few years, even after April 1884 when T.R. Keith & Co., an enterprising seed company based in Winnipeg, began to advertise in the *Bulletin*. That same year, the Renfrew Fruit and Floral Company, based in Arnprior, Ontario, made a late appearance on the Edmonton scene; by March 1885, it was just one of several companies advertising a variety of plant material for home gardens, including strawberry plants and roses. In 1890, by which time the population of Edmonton had risen to approximately five hundred and the hamlet was described as having a full range of services, including "six large mercantile establishments, whose stock contains everything from sides of bacon to ostrich plumes," a call went out for "50,000 farmers and their Families" to come and see the "Garden of the Northwest Territories."[31] Which garden, one is tempted to ask?

The decade of the 1880s saw the emergence of a new vision of nature, a nature transformed by trees, shrubs, fruits, and flowers that, while not native to the Edmonton area, would thrive in its soil and climate. From the preoccupation to bring this vision about, a process of change was set in motion that no subsequent efforts to check or reverse could stop.

In November 1883, by which time settlers were already experimenting with trees, shrubs, fruits, and vegetables for which they had managed to obtain seeds, seedlings, or rootstock, the *Bulletin* reported on evidence given to a House of Commons committee on immigration and colonization in northwestern Canada by botanist John Macoun.

Macoun, a dominion botanist with the Geological Survey of Canada, recommended species that could be expected to grow in new settlements such as existed at Edmonton, and his advice that "ash, soft maple, rim ash and oak will grow all over the prairie region of the North-West if propagated" was eagerly taken up. "It is not the intense cold that kills trees, it is their inability to resist great changes of temperature," he said, correctly predicting the experience of settlers who found greater dieback in their apple trees after a winter of variable temperatures.[32] As for apples, he advised against looking to eastern sources and recommended Russia instead, a recommendation that amateur and professional horticulturists followed despite the sometimes-disappointing results. In 1885, when the Edmonton Agricultural Society took up the cause of apple growing, an article appeared in the *Bulletin* urging governments at both the territorial and the federal levels to take on the cause.

> It is well known that in Europe and Asia, in districts of similar climate...apples and even pears are grown successfully. This being the case, there is no reason why the same varieties of fruit would not grow in the North West...[I]t would be both proper and possible for the government to send an agent to Russia or Siberia to make enquiries, and procure specimens of trees for experiment. If the experiments gave a reasonable probability of success, it would then be worth while to establish a permanent agency in Russia for procuring cuttings of trees, and one or more nurseries in the North West, in which the trees might be acclimatized under skilled care before being distributed.[33]

In the spring of 1886, after two years of study and deliberation, "An Act Respecting Experimental Farm Stations" was passed by the federal government, bringing into immediate existence the Central Experimental Farm in Ottawa under Dr. William Saunders and leading to the founding of four federal research stations in Alberta: Lethbridge (1906); Lacombe (1907); Fort Vermilion (1908); and Beaverlodge (1916). Edmontonians, represented by their agricultural society and by their board of trade, had lobbied unsuccessfully in 1890 for an Edmonton location.[34] Nevertheless, from 1886 onwards, the Dominion Government's contribution to the development of plant material for the West, combined with its practice of disseminating information on new varieties and proper cultivation techniques through local newspapers such as the *Bulletin*, did much to hasten the process of agricultural, arboricultural, and horticultural transformation in the Edmonton area.[35]

Enterprising Edmontonians, spurred on by the work of the Central Experimental Farm and by local promoters such as Oliver, launched into experiments of their own. In 1888, for instance, gardener and honey producer J. Knowles was reported as having a garden full of non-native species, including locust, catalpa and Norway spruce trees, golden willow that he "received from the east," and several varieties of cultivated red and black currants and gooseberries. He had planted tobacco for the second year in a row with somewhat less than convincing results and he had a "magnificent array" of flowers.[36] In 1889, Donald Ross made sure to make known his success with the strawberry hybrids he had obtained from the Central Experimental Farm and the following year he showed "a very fine sample of very large peas" in the fall agricultural exhibition. "This is a crop which has received too little attention in the past," reported the *Bulletin*, before concluding that "[i]t would add greatly to the wealth of the country if it could be made a success."[37]

The 1890s saw the introduction of Edmonton's first greenhouses, an increasing interest in varietal performance, and the creation of gardens that became showplaces for prospective settlers. In September 1891, the *Bulletin* reported that "[v]isitors to Edmonton this season have been astonished and pleased at the number and luxuriance of the flower gardens in town." Perhaps the best of these, the article went on to say, was that of crown timber agent Thomas Anderson, who was growing "[s]tock of all varieties, phlox, pansey [*sic*], pink, petunia, aster, candy tuft, sweet pea, daisy, verbena, marigold, [and] zinnia."[38] The process of transforming the garden, begun with some trepidation a decade earlier, was now confidently assumed to be inevitable, a matter of time and application to the cause.

The vision of a nature enhanced and ultimately transformed by introduced species and by newly developed hybrids was never intended to exclude entirely its cultural predecessor. Why might nature's garden not flourish in the same space as the domesticated garden? No fears of damaging ecosystems interfered with the vision, held by many Edmontonians, that nature's garden could exist in peaceful and complementary coexistence with the domesticated garden. From the beginning of the settlement period, there were gardeners who recognized the advantages of accepting what nature had to offer and including these offerings in their garden plans.

The notion that native berry fruits could be successfully integrated into the domesticated garden, thus relieving the gardener of the need to gather them from the wild, enhanced their appeal to many settlers. "Blueberries and cranberries, strawberries and raspberries, gooseberries and currants, are to be had for the picking," announced an article in the *Bulletin* in the fall of 1883, "but no doubt a rich reward not only in convenience of picking but in size and flavour of fruit is in store for the man who begins to cultivate them."[39] Five years later, with tree fruits still proving to be elusive, a long editorial urged farmers to

> transplant these small fruits, for your own benefit in the first place and for the public benefit in the next. By beautifying your own places you beautify the country, by supplying your own tables with home grown instead of imported fruit you save your own money, and prevent the amount of money from being sent out of the country. By cheapening and making more plentiful the delicacies accessible to the farmer you add to the attractiveness of farm life; and in many ways the country is made more pleasant to those who are here and more desirable to those who may wish to come.[40]

In 1889, the St. Albert Mission convent garden contained a combination of domesticated and native species growing together to create a pleasing whole. Imported Manitoba maples had been pruned carefully and were approximately twelve feet in height. Native saskatoon bushes planted in the same garden had also been pruned to make "a very pretty ornamental tree."[41] While we do not know if the sisters valued the saskatoon as much for its fruit as for its ornamental qualities, there is evidence that native fruits grown in domestic gardens were often judged to be superior to introduced varieties. In 1891, Douglas Petrie was reported as preferring the wild black currants he had transplanted to his garden over the cultivated variety. "The fruit is very large and of fine flavor, excelling the ordinary cultivated black currant both in size and quantity."[42]

Good growing weather helped to create a vision of nature in which both domesticated and native plants added to the bounty and beauty of the countryside. Such weather characterized the 1894 growing season. Ample spring rainfall and warm temperatures were followed in July by bright, sunny weather, creating what the *Bulletin* referred to in an editorial as a "Glorious Summer." With cultivated crops and uncultivated fruits and flowers growing equally well, the editorialist gave way to a rhapsodic description of the garden-like characteristics of the countryside:

> *The appearance of the country at the present time is such as must strike the dullest eye. The deep dark green of the poplar woods, the tall waving grass of the prairies, filled with flowers of a hundred beautiful varieties from the stately orange lily to the tiny white speck hiding close to the ground; the smooth undulating surface showing blue slopes and ridges in the far distance, but cut in two by the deep valley of the Saskatchewan and its wide swift flowing stream—it is the very model of a region in which use and beauty are combined in the very highest degree to make it desirable as an abode of man.*[43]

The integrated garden, based on the progressivist notion that, far from being compromised, nature stands only to be enhanced through cultivation and the introduction of new plant material, quickly became, and has remained, the dominant model for the home garden in Edmonton. It is also the model that informed the emergence of a parks philosophy and implementation plan during the first two decades of the twentieth century.

Although a small inventory of city parkland existed in Edmonton as early as 1906, it was not until 1907, when Frederick G. Todd presented his "Report on Parks and Boulevards," that a philosophy for parks development began to emerge. Todd believed that if people in cities "are to live in health and happiness, [they] must have space for the enjoyment of... nature...because it is the opposite of all that is sordid and artificial in our lives [and] is so wonderfully refreshing to the tired souls of city dwellers."[44] What he meant by the word "nature" can be inferred from his recommendations for particular parks.

For Todd, preserving nature implied some level of intervention. Rat Creek Ravine, for example, located just east of the downtown and held up by Todd as one of the most beautiful ravines in the city, would best be preserved by "keeping it as natural as possible," a preservation exercise that would require "carrying out a considerable amount of forestry work." "In some parts," he advised, "it would improve the present growth to take out a considerable amount of native poplars" and to replace them with "interesting trees" such as American mountain ash and evergreens of various sorts. Low-growing shrubs such as "high bush cranberry, Siberian dogwood, Siberian crab, and Japanese roses, all of which are hardy growers...would help to hold the banks."[45]

Morell & Nichols, the landscape architecture firm hired by the city in 1912 to produce a report on how to move forward with city planning, took to a new level of detail Todd's idea of an expansive river valley parks system linked by boulevards and supplemented with small and larger parks serving neighbourhoods and regions throughout the city.[46] The report emphasized its authors' belief that a city "can have no greater asset than the possession of a well planned system of parks, playgrounds, etc., where the public can enjoy the beauty of Nature and gain the recreation which is only to be found in the open air among the trees and flowers," although, as for Todd before them, the beauty of nature did not exclude features such as pathways, playgrounds, water features, and the introduction of non-native plantings.[47] In researching and preparing its report, Morell & Nichols worked closely with Paul von Aueberg, Edmonton's superintendent of parks, who, for one brief year, was empowered to translate ideas into action.

Von Aueberg, who had been a resource to Morell & Nichols during the fieldwork phase of its work and who endorsed the report's recommendations wholeheartedly, could not have imagined that his greatest opportunity for influencing parks development in Edmonton would be limited to little more than a year. In April 1912, he was put in charge of the newly created Parks Department with a staff, a substantial budget, and a mandate to acquire and develop parkland. Fifteen months later, in response to an unanticipated drop in city revenues, the entire department was let go and von Aueberg's many initiatives were put on hold. Not until 1947, two years after the end of the Second World War, was another specifically designated parks department created, leaving the maintenance and development of parks to languish (or not) in the hands of the City Engineer's Department in the interim. Nevertheless, during von Aueberg's short tenure as superintendent of parks, he set a course from which the city has never significantly deviated.[48]

On the policy front, it was during von Aueberg's tenure that Edmonton City Council made a commitment in principle to expand the inventory of ravine and river valley property, a principle that has been upheld over the years, resulting in a continuous stretch of linked parks and trails running throughout Edmonton's North Saskatchewan River valley and ravine system. It is a system currently managed according to guidelines set out in the *North Saskatchewan River Valley Area Redevelopment Plan* (1985), the *Ribbon of Green Concept Plan* (1990), and the *Urban Parks Management Plan* (2006–2016).[49] However, it was von Aueberg's approach to developing the parks already in his inventory that offers the best insights into the transformative vision of nature he wished to effect.

The small but strategically located Groat Ravine Park, located on 25.3 acres of land that fell victim in later years to the building of Groat Road, was featured in von Aueberg's 1912 end-of-year report. The land was assembled between 1910 and 1912, beginning with about seventeen acres of donated land that was rounded out with an 8.2-acre acquisition

Ladies walking in Groat Ravine Park, Edmonton, 1913. [GA NA-1328-6473]

costing the city just over twenty-seven thousand dollars. Both Groat Ravine and West End parks were accessible to those living in the downtown area of Edmonton and very close to the fashionable and newly developing areas of Groat Estate and Westmount. According to von Aueberg's report,

> Groat Ravine Park...was cleaned of dead and fallen timber early in the season, footpaths were constructed on the slopes of the ravine, and a caretaker installed, under whose supervision people are allowed to build fires and make tea at certain places. This little ravine was...extremely popular, as many as four hundred people being there on a Sunday, and it has now been supplied with a much desired extension to the north-west permitting the construction of a bridle path towards West End Park.[50]

Under von Aueberg, paths were carved out in existing parks, lawns were created to facilitate sports and picnics, a cricket pitch, a baseball ground, a children's playground, and one hundred benches were established in East End (now Borden) Park, while West End Park "was provided with a network of driveways and with drainage for some large sloughs, which it is proposed to transform into meadows."[51]

Von Aueberg's goal was to make parks accessible to the public and to provide the kinds of facilities in them that would encourage and support their use. To this end, he levelled land, filled up swamps, removed and pruned existing trees, laid out athletic fields, and built roads. Snow cover in winter prevented some of his improvements from being seen, he wrote, "and yet it is essential for the transformation of sloughs, wild grass land and thickets, into Parks."[52] That von Aueberg succeeded in accomplishing at least part of what he set out to do is made evident in the 1913 issue of *Town Topics*, in which the city is complimented for its parkland acquisition and development. "One park that is worthy of mention is Riverside [now Queen Elizabeth] Park," wrote the journalist, because "[h]ere one may be almost in the centre of the city and yet be in the heart of nature."[53] Clearly, from a public perspective, von Aueberg had struck the right balance between preservation and transformation in his approach to parks development.

A century after the emergence of Edmonton's first parks department, and in spite of the vast increase in parkland inventory and the ever more complex urban context of which it is a part, the city's parks philosophy has scarcely changed, although its approach to developing parklands has been modified. "As Edmonton grows and evolves so do its natural spaces," writes Martina Gardiner in an article about Borden Park: "Often there is a need to protect; other times there is a need to revitalize and expand."[54] Contemporary understandings of the concepts of protection and transformation have altered somewhat over time, giving rise, for example, to a new openness to "Natural Area Parks," which are "intended to conserve sustainable elements of our natural heritages."[55] Nevertheless, finding an appropriate balance between protection and transformation is one of the threads that ties Edmonton's contemporary gardeners and planners to their horticultural predecessors.

East End Park (Borden Park), Edmonton, 1913. [CEA EA-10-2926.4]

Gladys Reeves, Edmonton photographer, feminist, gardening advocate, and tree planter, was one such predecessor whose idiosyncratic conception of the relationship between nature and the garden informed her gardening advocacy. Reeves, an active member of the Edmonton Horticultural Society throughout her adult life, was both an advocate for the transformative benefits of gardening and one of the first to object when, in her view, Edmontonians failed to honour or properly steward the "natural beauty" of Edmonton's ravines and riverbanks. In undated notes for a speech she gave to the Edmonton Optimist Club, she referred to the ravine and riverbank system as a "gift from nature [given to us] to improve or to mar." In Reeves's view, "we have done more to mar them than improve them."[56] It was an opinion she voiced from time to time with reference to any development or city beautification initiative that, in her view, privileged transformation over protection.

The fate of Rat Creek (now Kinnaird) Ravine, the same one referred to by Todd and described as one of the most beautiful in the city, was of particular concern to Reeves. Her family had moved in 1912 from 101st Street, downtown, to a house in what is now the Cromdale area of the city on 112th Avenue near 79th Street. In later years, she owned a house in the same area. When, in 1931, the city was faced with having to replace a bridge over the ravine where it intersected 82nd Street, the Town Planning Commission recommended expropriating six houses west of 82nd Street to expand an existing park and replacing the unsafe wooden bridge with one made of steel and concrete. The Cromdale Community League lobbied to simply fill in the ravine. Reeves expressed her own views on the subject. The decision was a bad one, she argued, not only for the loss it would represent to nearby residents, many of whom had moved to the area for its proximity to a spot of nature, but also for the faulty logic used to justify the project. "If the desire to find work for the unemployed is the real reason for the [project]," she reasoned, "why not use the unemployed to preserve and beautify—not to destroy."[57]

Reeves's love of trees and commitment to city beautification led, in 1924, to her assuming the role of secretary for the newly formed Edmonton Tree Planting Committee. In the first year of its existence, the committee planted "700 young birch trees," which its members dug by hand from nearby woodlands.[58] The practice of using native trees persisted for a couple of years, but in 1926, due in part to the increasing difficulty of finding suitable sources of native species, the committee "obtained the use of one of the City's outlying park areas for the purpose of growing nursery stock for public planting." By the spring of 1927, when the committee applied to the city for a two hundred dollar grant, it had more than thirty thousand seedlings planted with species that were considered more suitable for the purpose, species such as American elm, green ash, mountain ash, oak, cedar, and Ontario maple.[59] Reflecting on this experience a few years later in notes for one of her many speeches, Reeves recounted criticism the committee had received for planting only native trees in 1924, but, with characteristic

Members of the Edmonton Tree Planting Committee at work, 1928. [PAA A7397]

stubbornness, she insisted, "one of the prettiest Avenues in the City was an avenue of Birch Trees on 85th Ave."[60]

Reeves's respect for, and love of, the natural world surfaced frequently to moderate and inform the zeal with which she advocated for city beautification. In this she resembled her friend and horticultural society colleague, University of Alberta Professor George Harcourt. Harcourt, who worked with Reeves on the very successful vacant lot gardening program in Edmonton, was as ardent an advocate for city beautification as she, but, like her, his advocacy was tempered by a respect for nature. In "Beautifying the Home Grounds," a published version of a radio talk given for the Department of Extension at the University of Alberta and addressed to homeowners, Harcourt advised listeners to consider native tree and shrub varieties as an alternative to more showy introduced plants. The native plants "will not be as rich in color and variety as when some of the hardy cultivated varieties are used," he warned, but "it is not so much what is planted as how and where it is planted."[61]

It was to be many years after the disappearance of both Reeves and Harcourt from Edmonton's horticultural scene that more radical understandings of the concepts of preservation and protection began to make their way into the horticultural consciousness of the city. Would Paul von Aueberg have so eagerly eradicated sloughs, grasslands, and thickets in the process of creating city parks, or might he have preserved some of them to create one of the natural area parks called for in the City of Edmonton's *Urban Parks Management Plan: 2006–2016*?[62] Not until the mid-to-late 1990s, during a long and ultimately unsuccessful struggle on the part of city conservationists to preserve an approximately eighteen-hectare site known as Little Mountain from development, did the need to protect entire ecosystems make its way to the forefront of public consciousness.

Calls by the Edmonton Nature Club to preserve a remnant piece of aspen parkland adjacent to the Little Mountain Cemetery in the northeastern sector of the city began when it was listed in the city's *Inventory of Environmentally Sensitive and Significant Natural Areas* (1993) and deemed worthy of protection.[63] In 1994, after a guided tour by Patsy Cotterill from the Edmonton Nature Club, Joy and Cam Finlay wrote an article about Little Mountain for the *Edmonton Journal* describing the site, explaining how it managed to survive undisturbed into the 1990s, and calling for it to be turned into a nature reserve along the lines of Winnipeg's Living Prairie Museum.[64] In late 1998 and throughout 1999, the impending loss of the Little Mountain natural area to development provoked organized protest from individuals and groups committed to the conservation of unique ecosystems. Ultimately, despite the persistent and resourceful advocacy of Cotterill and others, advocacy that won the cause strong support from some members of the Edmonton City Council, and despite sustained journalistic support from *Journal* columnist Ron Chalmers and *Journal* staff writer Allan Chambers, Little Mountain was lost in November 1999, when council reached the "joyless decision" to allow the property's owners to begin development.[65] Gone forever was "an area as large as several city blocks with unusual hydrogeology that produces a saline meadow plus bushes, balsam and aspen trees, and native grass growing from sod that's never felt the plow."[66] Gained, if not immediately, was political and public support for the notion that preserving nature's garden is not as simple as planting a tree; it involves protecting entire ecosystems. Five days after Chalmers's recording of council's "joyless decision," an editorial in the *Journal* called for a plan for the city's many "natural areas," noting that while the loss of Little Mountain as a natural area had become a lamentable fact, "it would be even more tragic if the city let this opportunity to establish a coherent policy on natural areas pass it by."[67]

The Little Mountain affair led eventually to city council's decision in 2001 to hire Grant Pearsell as its first conservation co-ordinator. In short order, this single position evolved to become the Office of Natural Areas, with Pearsell as director. With a small staff of four, a broad cross-departmental mandate, and a line of credit with which to purchase unique areas in need of protection, the Office of Natural Areas was given a stronger civic voice on issues of urban environmental protection. In 2006, a nonprofit organization known as the Edmonton and Area Land Trust was founded by six organizations, including the City of Edmonton and the Edmonton Nature Club. Dedicated to "conserving the natural heritage of the region through private stewardship," it launched immediately into a program intended to "conserve, protect and restore natural areas of ecological significance in Edmonton and the surrounding area." By 2012, the Office of Natural Areas had become the Office of Biodiversity, with a staff of nine and an added responsibility for parks policy. By 2014, it had grown again to include responsibility for parks planning.[68]

The Edmonton Native Plant Group, formed in 1997 as the Edmonton Naturalization Group, and perhaps the first Edmonton gardening organization to truly understand the importance of conserving native plants and, by extension, the ecosystems of which they are a part, acted quickly to salvage some of the ancient prairie grassland at Little Mountain. In the spring of 2000, despite efforts by the site's owner the previous fall to burn the vegetation, including the stands of aspen and balsam poplar, the group intervened. Led by

Cotterill, several of its members attempted to transplant a section of the native prairie to a safe site adjacent to Edmonton's waste management facility at Clover Bar, just east of the city, cutting the prairie as though it were turf, rolling it up and unrolling it at the transplant site. Although the transplant was not ultimately successful, the effort involved made the point that, whether we like it or not, "we must cultivate what we once took for granted." Close to a century and a half after Edmontonians first sought to transform the natural world they encountered to something that more closely resembled their concept of the garden, that same natural world, rapidly vanishing, has been revalidated and, in some measure, restored to prominence in the gardens and public parkland of the city.[69]

4 Among the First

More and more do I appreciate and marvel at the contributions that have been made, and particularly by the many individuals whose vocation was not horticulture, but through their love of the beauties of nature, their tremendous powers of observation and the challenge that their surroundings afforded, they have given to us a plant heritage that we can ill afford to forget.

—H.T. ALLEN

People who live in a place and learn to call it home are too vigilant about that place to allow it to pass out of their control.

—DAVID CARPENTER, *Writing Home*

GARDENERS are, by definition, innovators, always trying new seeds, new combinations of plants, new cultivation techniques, new experiments in the garden. But four men, contemporaries whose worlds intersected without overlapping, warrant singling out as having been among the first in the Edmonton area to step away from amateur putterings in the garden to embark on grand-scale horticultural projects. In 1905, Ontario-born Walter Ramsay resigned his position as principal of the newly opened Queen's Avenue School to launch a major greenhouse operation and florist business.[1] That same year, Georges Bugnet came to Canada from France to homestead, but he made his mark not as a pioneer farmer but rather as an author and plant hybridist.[2] Alfred Pike moved to Edmonton from Nottingham, England, in 1910 to work for A.E. Potter in the seed business, so perhaps it is not surprising that, five years later, he left to found his own seed business. However, Pike did more than sell seeds; he was the first local seedsman to engage with the horticultural community and apply himself to developing products that were uniquely suited to Edmonton's growing conditions.[3] By 1919, when sixteen-year-old Robert Simonet arrived in Edmonton, gardening was a popular, even a patriotic, pursuit. Simonet's drive to break new ground as a plant breeder, fuelled by a disciplined capacity to learn on his own, led eventually to the development of flowers, fruits, and vegetables that made their way into gardens everywhere, especially in Edmonton.[4] Ramsay, Bugnet, Pike, and Simonet are certainly not the only Edmontonians to have taken on horticultural challenges that presented themselves in a young and northern city, but they are among the first to do so here. And they are worth remembering, not just for what they accomplished as individuals but because their projects were so intricately connected with the place Edmonton was and went on to become.

Ramsay was one of the first in Edmonton to create a large retail florist business, one based initially on what he grew in his own greenhouses. Such businesses, each of which develops a personality differentiating it from its competitors, attract loyal followings and become "go to" places for decades. Some, such as the present-day Apache Seeds, Enjoy Centre, Greenland Garden Centre, Salisbury Greenhouse, and Wallish Greenhouses, remain in a family for generations. They set standards in aspects of horticulture such as floral design, plant choice, and landscape design. Often located at the periphery of densely populated areas, they function together to create a particular geography of the city, a geography well known to gardeners. Without them, Edmonton would be an unimaginably different place, unrecognizable to eyes accustomed to the colours, textures, and plantings found in contemporary gardens and landscaped areas. Ramsay, whose greenhouse and florist business burst upon the Edmonton scene in 1906, awoke the city to a whole new set of floral imaginings and expectations.

It took Ramsay a mere few years to discover that his passion for the "culture of beauty" could become a life work.[5] Born on a farm near Hamilton, Ontario, Ramsay was a twenty-seven-year-old schoolteacher with seven years of experience when he decided, in 1897, to move west. After a year spent teaching in Clover Bar, just outside Edmonton, he accepted a position with Edmonton Public Schools and took steps to settle down, marrying in 1901 and purchasing a house downtown on Elizabeth Avenue near Howard (102nd Avenue and 100A Street, on the site of the present-day Edmonton Centre). But Elizabeth Avenue was not to be his permanent address and teaching was not to be his career. Within a few years, Ramsay was to leave the certainty of teaching behind for the uncertainties of a business venture, the scale of which was something new for Edmonton.

Edmontonians were quick to notice that their teacher's talents and interests extended beyond the classroom. He had taken the somewhat unusual step of constructing a small greenhouse, thirty-five feet by eighteen feet, behind his Elizabeth Avenue home. Gladys Reeves noticed it in 1904, shortly after she arrived in Edmonton with her family and began to attend Queen's Avenue School. Years later, she wrote that she liked to walk by the Ramsay house on her way to school, stopping to admire the flowers in his greenhouse.[6] Reeves was likely a student at Queen's Avenue School the year Ramsay persuaded the board of trustees to beautify the school grounds by planting one thousand tulips there, a project that may have informed Reeves's own later interest in school ground beautification.[7] And Reeves was not the only one to take inspiration from Ramsay's garden. According to several accounts, the Elizabeth Avenue residence drew many local admirers, some of whom inquired about purchasing cut flowers to take home. Ramsay, thus alerted to the business opportunities awaiting him, did not dither; he resigned his school principalship at the end of 1905 to plan and open his own greenhouse and florist business.[8]

It may have been a bold move in 1906 to build a greenhouse complex in Edmonton for the express purpose of cultivating flowers and decorative plants, but it was not a rash one. Ramsay's sense of timing was perfect and he had prepared well. Edmonton was growing rapidly and consumer markets were opening up in items formerly considered to be luxuries. Ramsay, consciously or instinctively, decided to tap into this market. In December 1905, almost a year before the business was officially launched, he placed an advertisement in the *Edmonton Bulletin* announcing that he had cut flowers, flowering plants, palms, and ferns for sale to "[b]righten up your home for Christmas."[9] All sales were made from the Elizabeth Avenue greenhouse. Then, during the winter of 1906, he travelled east to examine large operations such as Dale Estates Nurseries in Brampton, Ontario. In May 1906, when Ramsay filed building plans with the City of Edmonton, the *Bulletin* aroused anticipation with its report that "[t]he building he intends to put up will have 150 feet frontage on Eleventh street, covering the entire frontage of the three lots and will extend 110 feet back, making one of the largest greenhouses west of Winnipeg."[10]

Ramsay's greenhouses were located on Victoria Avenue (100th Avenue) between 110th and 111th streets, just east of the General Hospital. Even in 1906 it was a fairly central location and thus rapidly became both a city landmark and a destination for an outing. The first five "King patent" greenhouses, costing an impressive ten thousand dollars, were up and ready for the grand opening at Thanksgiving, 1906.[11] Four more were added the following summer, to make a total of nine. Nine greenhouses, ranged side by side, made a monumental and impressive sight, photographs of which were often featured on the firm's advertisements. Built into the southeast corner of the complex was a small office from which, until 1915, all sales were handled. And, across 110th Street to the east, on a large double lot, Ramsay built a

Ramsay's greenhouses, Edmonton, 1914. [GA NC-6-902]

new home for himself and his family, one that was even more impressively landscaped than the Elizabeth Avenue residence and one that had room for a big vegetable garden and a 57' x 12' hobby greenhouse.[12]

Three days before the impressive opening, for which Irving's Orchestra played between 2:30 and 6:00 P.M., the *Bulletin* ran an anticipatory article describing the delights that awaited all who planned to attend; they would see "beds of roses, carnations and chrysanthemums," "hundreds of pots of crimson geraniums," and "miniature forests of palms and ferns" that showed to advantage in the "warm autumn sunlight." French and English violets and a new strain of sweet peas rounded out the flower section and, for those interested in extending the vegetable season, twenty-five thousand lettuce and three thousand tomato plants had been planted. Thus, in one fell swoop, Ramsay moved Edmonton forward on the scale of horticultural sophistication and himself into position as one of the city's most respected businessmen.¹³

Between 1906 and 1915, while sales were being handled from the small office attached to the greenhouse complex, all plant material sold was grown on site. As many as five of the nine greenhouses were given over to roses and carnations, which were reputed to have been Ramsay's favourite flowers. Among the varieties of rose grown for the 1906 opening was a pink variety from Dale Nurseries that went by the name of 'New Canadian Queen'. Tea roses destined for bridal bouquets were appropriately named 'Bride' (white) and 'Bridesmaid' (pink). Chrysanthemums, including a Japanese variety called 'Golden Wedding', were available that first autumn. As the seasons progressed, plantings inside the greenhouses shifted from bulbs to bedding plants and then to houseplants. Three empty lots adjacent to the greenhouses and fronting Jasper Avenue were conscripted in the early years in order to sell bedding plants from wooden flats. And, during the summer, the gardeners on staff had time to design and plant gardens for individuals and for organizations such as the Masonic Hall downtown, of which Ramsay was a member.¹⁴

The fifteen-year period that began in 1914 with the outbreak of the First World War, and ended in 1929 with the stock market crash followed by the Great Depression, was a period of consolidation, expansion, and adaptation for Ramsay. The greenhouse complex, which seemed to move closer to the centre of the city as residential areas expanded around it, maintained its iconic presence on Victoria Avenue. This presence was indelibly associated with the reputation and status of its proprietor, whose many contributions to the community extended to a stint on the Public School Board, including two years as chair, as well as active participation in the work of his church and service organizations. In 1915, however, in a move that paved the way for a separation of the retail and wholesale sides of the business, Ramsay opened his first store in downtown Edmonton and began to order some stock from outside suppliers. Then, in 1921, the first in the city to recognize a market for transnational and international deliveries, he joined the American organization FTDA (now Florists

Seeds and Plants
FOR YOUR GARDEN

MANY OF THE PRIZES LAST YEAR WERE WON BY PURCHASERS OF OUR SEEDS

It's the Quality that Counts

Agents for Carter's Tested Seeds, as well as for other Choice Strains.

WINDOW BOXES AND BASKETS FILLED AT LOWEST PRICES

Bulbs for Spring Planting

GLADIOLI, DAHLIAS, CANNAS, PEONIES SHRUBS, PERENNIALS

FERTILIZERS—Bone Meal, Nitrate of Soda, Potash, etc.
CIPPS—The Wonderful Plant Food in Tablets easy to apply

WALTER RAMSAY LIMITED

EDMONTON'S LEADING FLORISTS,

10218 JASPER AVE.   PHONES 23554 and 5535

Advertisement for Walter Ramsay Limited from the Edmonton Horticultural Society Prize List, 1926. [CEA MS 89]

Walter Ramsay Florist, Edmonton, June 12, 1942. [PAA Bl.390]

Transworld Delivery or FTD). According to McCracken, Ramsay later chaired an Alberta Branch of the American organization and became active, after its formation in 1943, with the Allied Florists and Growers of Canada.[15]

The Great Depression of the 1930s dealt Ramsay's business crippling blows, exacerbated by the burden of having to maintain the greenhouses. Business declined alongside the declining fortunes of Edmontonians, making it difficult for Ramsay to pay his staff. Then, during the winter of 1934–1935, the first five of his greenhouses collapsed. That the business survived must have been due to Ramsay's earlier perspicacity in creating a separate identity for his downtown store. Although the greenhouses were torn down in 1944, reclaimed by the city for back taxes, the retail side of the business was unaffected. Ramsay's reputation as a florist, a reputation that had been built around the production of his own greenhouses, endured through his own and his son's lifetimes, sustained by the pre-eminence of both men in the florist industry Ramsay had done so much to build.[16]

Ramsay's ability to anticipate trends and shape them to his own advantage, combined with an equal ability to adapt to changing economic and social conditions, contributed to his firm's longevity and reputation. His death in early December 1958, reported in the *Edmonton Journal* as the loss "of one whose unstinting labors did much to beautify the city and its daily life," was followed less than a year later by an article announcing that the family home on 100th Avenue was about to be torn down to build a modern apartment building.[17] Although Walter Ramsay Florist was carried on by its founder's son Donald until the latter's death in 1974, the shop disappeared entirely from the Edmonton scene in the early 1980s, leaving no physical trace of the grand project that had eased Edmontonians away from their dependence on home gardens and towards the notion that gracious, even sophisticated, living could be achieved year-round with the aid of plants and flowers purchased from the local florist.[18]

It was also in the early 1980s that another of Edmonton's long-standing horticultural institutions, Pike & Co. Ltd., was sold to a

Manitoba company. Pike Seeds, which opened in 1915 without the fanfare that had attended the 1906 opening of Ramsay Greenhouses, built a reputation equal to that of Ramsay but founded on a different premise, the premise that with the right seed and plant material, gardeners could grow their own flowers and vegetables with ever-improving results.

Obtaining a good supply of reliable seed was, in the early years of the Edmonton settlement, not always easy. The relative shortness of the growing season meant that many plants could not produce seed before frost arrived, including the potatoes, root vegetables, and cabbages that grew so well in settlers' gardens and quickly came to be considered staples. Ukrainian farmers, who valued cabbage above all vegetables, produced seed by storing some heads through the winter in basements or root cellars, replanting them in the spring and then harvesting the seed from the flower stalk that appeared during the second season.[19] For those who bought seed, either by mail order or from local outlets such as Brown and Curry, the Hudson's Bay Company, Ross Brothers, or A.E. Potter & Co., viability was always an issue. In 1915, when Pike opened a business that specialized in seed sales, he carved out a niche for it by ensuring that everything he sold would grow in local conditions. "Pike's Seeds are the finest in the world for this climate" became the company slogan, and it was one that ensured the popularity of the brand even after the company's 1982 sale to McKenzie Seeds of Brandon, Manitoba. A. Pike & Co. may have been the first and best known of the local seed companies, but it, along with companies formed later such as Apache Seeds, Robertson Seeds, and the Seed Centre, were often favoured by Edmontonians over companies based elsewhere, partly because they were owned and run by fellow Edmontonians who understood local growing conditions and could be relied upon to guarantee their products.[20]

Pike came to Edmonton from Nottingham, England, in 1910, having decided that his eyesight would not hold up to the legal career in which

> *Alfred Pike, n.d. [PAA PR1974.173/681]*

Edmonton's 100A Street looking south to Jasper Avenue with a view of Pike's Seeds, c. 1930. [CEA EA-275-1030]

he had begun to train. Fifteen years old, and already a keen gardener, he began to work in the seed sales division of A.E. Potter & Co., a company that established itself in Edmonton as a transfer company before branching out into "lines of industry" such as seeds.[21] Sensing both an opportunity and an aptitude for the business, Pike struck out on his own at the age of twenty to found A. Pike & Co., Seed Merchants. Thus began Pike's long career as Edmonton's best-known and respected seedsman, a career that ended only when he died on November 26, 1981, at the venerable age of ninety-six.[22]

Unlike Ramsay, who gave up his first plan to create a florist business based on locally grown plant material, Pike held to his original intent, sourcing and supplying an ever-increasing range and quality of product that could be reliably grown in Edmonton and northern Alberta. The two men were in tacit agreement, however, as to the importance of a downtown location. In 1915, the year Ramsay opened his downtown store, A. Pike & Co., Seed Merchants opened at 10049 Jasper Avenue, moving in 1945 to a larger but equally central spot at 10039–101A Street. Only in 1976, by which time a central downtown location was no longer a necessary corollary to forming a strong business identity, did Pike & Co. move out of the downtown, by which time its disappearance as an independent entity was imminent.

A keen gardener himself, with a particular interest in growing roses, Pike understood the need to ensure that what he sold would grow. A connection made with Professor George Harcourt, possibly in 1918 when both men were active with the Edmonton Horticultural Society, led to an arrangement that Pike, himself, described in a 1975 seed catalogue celebrating the sixtieth anniversary of the company. According to this arrangement, Pike secured seed samples from suppliers in Holland, England, France, and the United States, seeds that he then tested in a plot allocated to him for the purpose next to the University of Alberta Greenhouse. After selecting the hardiest varieties, Pike attempted to produce seed from them in Edmonton, resorting to the Okanagan region of British Columbia as a more feasible alternative. All seed, he claimed, was then tested for viability in germination machines before being sold. This arrangement, one of several efforts taken to ensure the quality of his stock, earned Pike a reputation for excellence that persisted.[23]

That Pike remained at the helm of his business until the time of his death in 1981 was doubtless due to a failure of succession. In 1953, A. Pike & Co., a single proprietor business, was reorganized as a limited company with two major shareholders, Alfred Pike Senior and Alfred Pike Junior, and renamed Pike & Co. Ltd. While Pike retained the position of president, with his son taking on the role of vice-president, the younger Pike, for reasons unknown, failed to establish himself as successor. When Alfred Pike Junior died in 1966, long-time business manager Alec Wilson and sales manager Walter Dorin became shareholders and partners. The elder Pike, together with Wilson and Dorin, celebrated the sixtieth anniversary of the company in 1975. It was Wilson and Dorin who, in 1982, arranged the sale of Pike Seeds to McKenzie Steele Briggs Seeds Ltd. of Brandon for a little over a million dollars.[24]

Throughout his business career, but particularly until the mid-1950s, Pike maintained strong connections with Edmonton's horticultural community, especially as that community was mediated through the activities and programs of the Edmonton Horticultural Society. In 1918, he joined the society's board of directors and remained on the board until 1936, assuming a variety of positions and responsibilities.[25] It was a period during which the society was particularly active in Edmonton, renting lots for gardens and monitoring rental agreements, staging annual flower and vegetable shows that sometimes drew thousands of spectators, assisting with city cleanup programs and tree-planting committees, modelling best gardening practices by setting up demonstration gardens, mounting floral displays in downtown shops and hotels to impress visitors, and sponsoring speakers. As a board member, Pike was involved with these programs, helping to organize events and offering lectures on a subject of particular interest to him—growing roses. In 1933, by which time Pike was involved with the manager of the Capitol Theatre, Walter Wilson, on a scheme to raise Edmonton's profile as a place where roses could be grown, he donated prize money to the horticultural society to support the creation of new categories for roses in its annual shows.[26]

For a city with the harsh winters and dry summers that are Edmonton's typical lot, the plan devised by Wilson and Pike to create a reputation for Edmonton as the "Rose City" of western Canada, a plan they presented as a campaign, was, at the very least, optimistic. However, if success is to be measured in terms other than those that make slogans, it worked. For twenty years, beginning in 1933 when the first Capitol City Rose Show was held, Wilson and Pike created and sustained an interest in growing roses within the gardening community and fostered an appreciation for roses in the general public that has probably never been surpassed in the city. It was a long-lasting partnership, with Wilson supplying much of the promotional effort and a venue for the show in the foyer of the Capitol Theatre and Pike organizing the supply and sale of roses to gardeners. Together, they wrote and supplied, free of charge, a booklet on rose growing, created the annual prize lists for the

PIKE'S SEEDS

The Finest Seeds in the world for this climate

We Specialize in Exhibition Strains of
VEGETABLE and FLOWER Seeds

"EVERYTHING FOR THE GARDEN"

Gladioli — Dahlias — Pæonies — Hardy Bushes — Hedges
Shrubs — Shade and Ornamental Trees
Fruit Trees and Bushes

SWEET PEA SEEDS

We have a full stock of Named Varieties of Sweet Pea Seeds to Produce Exhibition Blooms.

Marigold Colarette

LAWN GRASS SEED

The most beautiful Lawns in Edmonton are made with Pike's Lawn Grass Seed.

New this year:—
ANNUAL SCARLET DOUBLE HOLLYHOCK
Package 25c

ROSE BUSHES
15,000 Rose Bushes for Our Customers to Select From
All the Rose Bushes we list in our Catalogue are Budded on a Special Briar, which has been proven to be the best to withstand our severe climatic conditions.

Phone or Call for Our Catalogue

Phone 22766
PIKE & CO.
SEEDSMEN and FLORISTS
10049 Jasper Avenue
EDMONTON — ALBERTA

Advertisement from the Edmonton Horticultural Society Prize List, 1938. [CEA MS 89] shows, and obtained prizes and prize money to attract exhibitors. Because both men were keen rose growers, they made themselves available to answer questions put to them by new or experienced gardeners.[27]

Quantities of roses sold by Pike during the 1930s were impressive for a city of eighty to ninety thousand residents, as many as 3,500 bushes in 1933 and six thousand in 1934.[28] In the Edmonton Horticultural Society's annual prize list for 1938, Pike advertised having a stock of fifteen thousand rose bushes for sale, while a report in the *Bulletin* in June 1949 estimates that thirty-six thousand rose bushes were sold in the city that year.[29]

Pike's commitment to sourcing products that were adapted to Edmonton's climate and growing conditions was one of the reasons Walter Dorin, sales manager and part owner in the company, liked his job. According to Dorin, Pike dealt with suppliers from around the world, including Britain, Holland, Belgium, and the United States. Sweet peas, which were much grown in Edmonton in the 1930s and 1940s, always came from Carters in England. Carrot seed came from San Francisco. Roses, during Dorin's time with the company, came from Eddie's Nurseries in the Fraser Valley in British Columbia. The supply of corn, bean, pea, and radish seed was contracted to a nursery in Armstrong, British Columbia. Closer to home, Pike sustained his relationship with the Faculty of Agriculture at the University of Alberta and much of his seed was developed and tested there. Altagold corn, for instance, was advertised in the 1969 catalogue as having been "raised by the University of Alberta and introduced by us to the gardening public." According to Dorin, Pike sold about ten cultivars that were developed by Simonet. And, of course, as the series of Bugnet roses became available, they, too, were brought in and sold by Pike.

From the beginning to the end of his remarkable sixty-six-year career in Edmonton as a seedsman, Pike witnessed, and assisted in, a radical transformation of the gardening scene, due in large part to the dramatic increase in the range of hardy plant material made available to local gardeners. Pike's part in this transformation was to test, promote, and supply what others developed. He had been in business many years before he could begin to sell plants that had been developed locally, and it is impossible to guess the relative degree of excitement and satisfaction this development may have inspired. And for their part, plant hybridists like Bugnet and Simonet were lucky to have laboured at their hybridizing when the local appetite for their product was insatiable, and when the lines of communication between plant breeder and the gardening community were so various and relatively informal.

Edmonton's challenging climate has inspired a host of amateur gardeners to try plant breeding for themselves, to engage in the never-ending search for new varieties that marry perfection of form in a plant to what Bugnet called, in an article for the *American Rose Annual*, "total hardiness."[30] Farmers and gardeners, encouraged by articles in the *Bulletin*, carried out their own plant trials and their findings were sometimes reported in local papers. Government departments of agriculture, both federal and provincial, established stations throughout the country that were staffed and equipped to carry out research and to support research activities in their regions, including research undertaken by amateurs.[31] The Faculty of Agriculture at the University of Alberta was equally supportive of amateur efforts after its formation in 1915 and it offered amateurs the added resource of a university library. Ultimately, the accomplishments of self-taught breeders such as Bugnet, Simonet, and those who followed in their footsteps illustrate the years of exacting, tedious work required to expand the repertoire of plants available to local gardeners and farmers. Partly on account of the plants they registered to be sold on the market, and partly on account of their unregistered experiments that populate local gardens and are often integrated into the breeding programs of peers and successors, local plant breeders are invisible animators of the city's gardens.

"Now I give pleasure to thousands of women" Bugnet was recalled as having said to writer and translator David Carpenter about his 'Thérèse Bugnet' rose. Carpenter can still remember Bugnet's smile as he said this, and the gravelly voice of a then very old man.[32] Bugnet, who died at St. Albert, Alberta, in 1981, was too frail during the last years of his life to continue with the journalism, writing, and plant-breeding experiments that had brought him renown in local, national, and even international circles, but he was not too old, it seems, to have lost either his sense of humour or his wit. He ended his life as he had lived it, a man of humble means but possessed of a rich mental and spiritual life. "There is always inside of me," he was quoted in 1959 as having said, "that if I tried to turn my deeds into money something unpleasant would happen. On the other hand, as long as wealth comes in fame and in fun, it can apparently stream in endlessly."[33] Bugnet's many and various "deeds," which include but are certainly not limited to his rose hybrids, brought him their share of fun, if fun is to be measured in units of satisfaction and pride, and they continue to bring him fame long after his death. They remain as evidence of one pioneer's abiding passion for the place he chose to call home.

Bugnet was born at Chalon-sur-Saône, France, in 1879, but his family moved first to Beaune and then to Mâcon before settling in the Burgundian city of Dijon in 1894. Under the influence of his mother, who had determined that her eldest son would enter the priesthood, Bugnet enrolled in 1896 in the Grand Séminaire de Dijon. But although he was influenced by this phase of his education and remained within the fold of the Roman Catholic Church throughout his life, he was also an independent thinker who turned to more secular classical studies for intellectual nourishment. In 1899, he enrolled in the University of Dijon, served a compulsory year in the French army, and embarked on a study of arts and letters that he hoped would lead to a vocation as a university professor. In 1904, he accepted a position as editor-in-chief of *La Croix de Haute-Savoie*, a Catholic newspaper, and, in April of that year, married twenty-two-year-old Julia Ley of Dijon. It was while working as a journalist, said his biographer, Jean Papen, that he succumbed to the lure of the West as it was represented in the propaganda brochures advertising homesteads with excellent prospects for farmers.[34]

Georges and Julia Bugnet with Georges's brothers Maurice and Charles, Dijon, France, c. 1904. [PAA PR.1978.0087.2]

What induced Bugnet to leave his family, his country, and a job that appeared to be eminently suited to his talents for the unknown and probably unsuspected rigours of life as a farmer in the then very new province of Alberta? Perhaps he was eager to leave behind the influence of a domineering mother. Perhaps he was stifled or constrained by French society. Perhaps he was in search of adventure and freedom. Perhaps he thought he could make a fortune in short order and return to France a wealthy man. All of these reasons were advanced in writings by or about Bugnet at one time or another and all are likely to have influenced his decision to emigrate.[35] Not to be underrated, however, is the appeal made to the young Bugnet's imagination when, as an impressionable adolescent, he read about the Canadian northwest in a publication written by Oblate priest Émile Petitot.[36] Petitot travelled to the northwest of Canada in 1862, living among the Aboriginals, learning and recording their languages, and going on to write extensively about them.[37] According to Papen, it was Petitot who first aroused Bugnet's curiosity about the Canadian northwest and its Aboriginal inhabitants, rendering him open, when the opportunity presented itself, to the prospect of a homestead in Alberta.[38]

If Bugnet had harboured any thoughts of returning to France a rich man, it did not take him long to give them up. The Dominion Government "must have been quite an optimist," he said in an interview for a history of the Lac Ste. Anne district, to have represented homesteading in Alberta as a "get-rich-quick proposition."[39] By 1908, he was a Canadian citizen and in his later years he was known to assert that he had found "a broader outlook" in his adopted country than he could ever have found in the country of his birth.[40] He, alongside his wife, Julia, chose to be buried in the Catholic cemetery at Lac La Nonne, surrounded by the countryside that inspired both his literary and his horticultural works.

The Bugnets left France at the end of December 1904, disembarked at Saint John, New Brunswick, on January 5, 1905, and arrived at St. Boniface, Manitoba, their first destination, two days later. After a winter in Manitoba, where Julia gave birth to Charles, their first son, and Georges began to acquire some Canadian farming skills, they made their way to Alberta where Georges worked for St. Albert farmer Oscar Terreault. In March 1906, a little more than a year after their arrival in Canada, the Bugnets moved to their one-hundred-and-sixty-acre homestead, located near the present-day hamlet of Rich Valley in what is now the Lac Ste. Anne County.[41]

In September 1906, a few months after they had begun homesteading, Julia gave birth to a second son, Paul, who died as an infant in a fire, but the couple went on to have eight more children; in all, four sons and five daughters were raised on the Rich Valley farm.[42] Life on the farm was difficult, according to Bugnet, involving setbacks such as the loss of a small herd of cattle they had bought with the aid of a loan and the failure of an attempt at tree farming to yield acceptable returns.[43] On the other

> *Georges Bugnet, n.d. [PAA PR.1978.0087.1]*

hand, he enjoyed the solitude of the countryside and was stirred by a beauty in the landscape so different from anything he had known in France. It was here, particularly after his sons began to take on the farm work, that Bugnet carried out the majority of his hybridizing and it was here, according to his biographer, that his experience of nature stirred him most profoundly.[44] It was also here, some time before the outbreak of the First World War in 1914, that he was paid an extended visit by members of his family from Dijon, including his sister Thérèse, who arrived in 1908 and returned to France in 1914 to join the Carmelite Order in France.[45] It was this sister for whom he named his most famous rose.

In 1947, Bugnet sold all but approximately fifteen acres of his land in order to devote his time fully to writing and plant hybridizing. A year later, as he neared seventy years of age, he joked with a newspaper reporter that plant hybridists tend to live forever, kept alive by curiosity as they await the results of overlapping five-year experiments.[46] Not surprisingly, in 1954, when he sold the home grounds and test plots to the provincial government to be preserved as a historical plantation, Bugnet moved part of the breeding program with him to a residence in Legal, where he lived until after Julia's death in 1970.[47]

Bugnet lived out his life in a seniors' home in Legal, close to the countryside with which he identified so strongly. According to Carpenter, his "problem was that he could not die," never having had a sick day in his life.[48] In 1978, at the age of ninety-nine, he was awarded an honorary doctorate by the University of Alberta in recognition of both his scientific and literary achievements and, on January 11, 1981, a month short of his 102nd birthday, he took his final leave of the place he had done so much to build.[49]

If Bugnet's literary works demonstrate keen observational powers and an almost mystical appreciation for the complexity and beauty of nature, his work in horticulture suggests an equally keen desire to probe beneath the surface of natural phenomena. Papen links this scientific curiosity to his subject's deeply held religious beliefs, suggesting that it was through intimate and constant contact with the "secret and grandiose laws of nature" that Bugnet could best approach and understand the wisdom of the divine plan.[50] But Bugnet also understood the role played by the natural world in the everyday lives of individuals, and it was a perceived need for horticultural enrichment in the family's physical surroundings that induced him to begin his experiments. Understanding, as he did, that embellishment was a prerequisite to creating a true attachment to the land, he began almost immediately to search out plants that were hardy, decorative, and functional by writing to the federal research stations in Indian Head, Saskatchewan, and Brandon, Manitoba, for seed. Very little of the seed obtained from these sources grew, but, while watching and waiting, the seed of curiosity was firmly planted in Bugnet's head; it was at this point, some time around 1912, that he decided "to enter the game" and try plant breeding for himself.[51]

What dawned on Bugnet, and provoked his interest in the tedious work of plant breeding, was the realization that climatic conditions at his Rich Valley homestead were even harsher than those that existed at the various experimental stations from which he had obtained seed. "As a plant breeder," he is quoted as saying, "I thought, at first, our location not at all suitable, yet out of the very failure in those first attempts to grow 'hardy' plants, arose the discovery that we had been led to a most carefully selected spot to manufacture special stuff, possibly the hardiest in the world." So, with the help of the University of Alberta library, Bugnet launched into a study of "plant geography," gathering information that, supplemented by fifty-three-cent stamps, he used to request seed and cuttings from governments and institutions in Europe, America, and in northern countries such as Japan, Manchuria, Finland, and Russia. His efforts were amply compensated when he began to receive, "year after year," seeds from "the toughest parts of the world."[52]

Bugnet took his cue as a plant breeder from Luther Burbank, American nurseryman and plant breeder, who operated out of California, and N.E. Hansen, the renowned Danish American head of the Department of Horticulture and Forestry in South Dakota from 1895 to 1937.[53] Following Hansen's methods, and in order to keep as much space available for new testing as possible, Bugnet gave names to his most promising selections and hybrids and then sent them to federal and provincial research stations, as well as to individual nurserymen, to be tested. His goal, especially during the early years of his breeding program, was not so much to decorate the gardens of city dwellers as to support farmers, "who had no time to coddle tender plants" but who nevertheless wanted to live with good shelter belts, edible tree fruits, and hardy ornamentals. Professedly uninterested in receiving money from his "inventions," Bugnet left the business of registering plants to others.[54]

Not surprisingly, Bugnet's early recognition in the local horticultural community came not from a hybrid but from a selection of seed he had obtained from St. Petersburg, Russia, well before the outbreak of the Russian Revolution in 1917. *Pinus sylvestris* 'Ladoga', referred to by Alberta Agriculture and Rural Development as the Bugnet Scots pine, and often simply as the Ladoga pine, turned out to be something of a winner on Alberta farms when planted as shelter belt material; it was grown in large quantities for distribution at the Oliver Tree Nursery on the northeastern outskirts of Edmonton by the Alberta Department of Agriculture. In selecting it, Bugnet gained a great respect for the role played by climate in any breeding program and it was a respect he carried with him into his hybridizing program.[55]

Between April 1940 and February 1941, when he was beginning to achieve success with his rose hybrids, Bugnet corresponded with J. Horace McFarland, editor of the *American Rose Annual*, over an article Bugnet was writing for the 1941 edition of that publication. McFarland's letters to Bugnet reflect the former's restrained excitement as they discuss the role played by climate in the mating of cells in any breeding program. Bugnet's argument, which challenged the notion that genes are the sole determinant of a plant's characteristics, was summarized by McFarland as implying "that the climate under which the mating of the cells takes place has quite a bit of influence on the progeny." It was an argument McFarland resisted at first but began to find persuasive. He wrote, somewhat excitedly, to Bugnet on February 28, 1941, confessing that "[t]he possibility you suggest as to the use of frost when the mating is proceeding is fascinating." Although the correspondence ends here, leaving Bugnet's hypothesis in scientific limbo, it is certain that Bugnet himself never lost his interest in climate as a determinant in achieving the total hardiness he sought for his hybrids.[56]

When he began his breeding program, Bugnet looked to fill horticultural gaps identified by settlers in the northwest, including the lack of tree fruits such as apple, pear, and plum. These were the focus of his early experiments and they resulted in some promising introductions. However, although he introduced selections of caragana and hawthorn, a plum cultivar he named 'Claude Bugnet', and two honeysuckle cultivars that he named after himself and his wife, testing them as he did all his promising selections and hybrids at the federal government research stations in Morden, Manitoba, and Beaverlodge, Alberta, and the provincial station at Brooks, Alberta, roses came more and more to occupy his attention, especially after the mid-1920s.[57]

Bugnet's rose-breeding program began, as he explained in his 1941 article for the *American Rose Annual*, with a respect for what he called "the brave little wild rose." "In the middle of the summer the whole country, up to the Arctic seas, unfurls countless millions of rose buds, pale pink to bright red, each plant blossoming for two to three weeks; each plant a greater marvel than any invention of man."[58] It was with native prairie species, such as *Rosa woodsii* and *Rosa acicularis*, that he began the breeding program that resulted in a host of named roses, some of which were ultimately picked up by nurserymen and registered before making their way into gardens around the Prairies. Many of these roses were named after female members of his family: 'Thérèse', 'Marie', 'Martha', 'Louise', 'Madeleine', 'Rita', and 'Betty'. Two, 'Lac la Nonne' and 'Lac Majeau', were named after the two lakes near his homestead that were major reference points in his life and writings, and one was called 'Nipsya', the name of the central character in his 1929 novel of the same name. Of these, 'Marie Bugnet', a compact shrub rose with double white fragrant blooms, is still sold widely today, but it is the 'Thérèse Bugnet', with its masses of double pink blooms that begin to emerge in May and continue into September, that has earned him the most fame. Registered in 1950, with Bugnet's consent, by the Saskatchewan plant breeder and nurseryman Percy Wright, the 'Thérèse Bugnet' rose quickly became a staple in urban and rural gardens.[59]

The Second World War was well underway in 1941 when Bugnet published his article "The Search for Total Hardiness." A consciousness of the war as a backdrop to his research permeates the article. He thanks God for the fact that the "brave little wild rose" is "perfectly useless, so far, for war purposes," and he is cognizant of the fact that "in these years of terrible wars" he has had the privilege of devoting himself to "patient cooperation with that mysteriously and magnificently creative power which some call God, and some other Nature, meaning, after all, the same thing, the same unfathomable entity." In the end, Bugnet's grand horticultural project, which was to breed hardy plants that answered both practical and aesthetic needs within his own community, accrued meaning not just because he succeeded but because of how he succeeded. Bugnet's horticultural and literary works are products of the same striving mentality, which found a physical and spiritual home in the rugged countryside northwest of Edmonton.[60]

Bugnet was forty years old and well into his initial plant-breeding experiments when, in 1919, fellow countryman, sixteen-year-old Robert Simonet, arrived in Edmonton from France. The move from Europe to Canada was less complicated for Simonet than it had been for Bugnet and involved fewer permanent leave-takings. The First World War had just ended. One of Simonet's two sisters, just a year or so older than him, came to marry a Canadian soldier she had met in France during the war. Simonet accompanied her, drawn perhaps by the thought that he could make a future for himself more easily in Canada than in France. That he would go on to become an internationally recognized plant hybridist probably never occurred to him, although he had already acquired a keen interest in gardening and plants from the grandmother who raised him. An intense but shy man who loved to talk with fellow enthusiasts about his interest in plant genetics,

Simonet was once described by someone who knew him as "the most thoroughly grounded uneducated geneticist I have ever known and also the most modest."[61] While the modesty may have been an attribute of character, the thorough grounding in plant genetics came through hard work and application.

Within a decade of his arrival in Edmonton, Simonet had manoeuvred himself into a position from which he could begin his plant-breeding experiments. By working at a series of jobs, including a period at the Misericordia Hospital as a boiler stoker/gardener and another with market gardener Peter Juchli, Simonet was able to support himself, accumulate some financial resources, and become acquainted with the particular challenges of gardening in Edmonton. At the same time, he made use of both the Edmonton Public and the University of Alberta libraries to study the principles of plant genetics and to make his first contacts with professors of horticulture in the Faculty of Agriculture. It is even possible, according to his niece Janine Dunn, that Simonet launched a trial market garden business and began his first experiments in plant breeding in the 1920s on land he rented from the city.

In 1929, Simonet married Lillian Herard of Edmonton, and in 1930, he bought five acres of land for a market garden in King Edward Park, at 7529–81st Street, where they remained until 1957 when they yielded to development pressure and moved to a larger acreage near Sherwood Park. Throughout the 1930s and into the 1940s, the Simonets earned their living by market gardening while Robert continued to work on the experiments that eventually led to his first big breakthrough. Lillian was, according to all who knew her, a hard worker who could do any of the heavy tasks that needed to be done. Simonet's nieces and nephews helped out, and even his mother, who followed her three children to Edmonton, joined in the family enterprise by fronting sales each Saturday at the City of Edmonton Market. According to Dunn, there was also a spot on the payroll for any neighbourhood child who expressed an interest in greenhouse and garden work. But for Simonet, who disliked selling and whose interests were constantly roving beyond mere production, market gardening was simply a means to an end. Ironically, the

Robert Simonet at work in the greenhouse, c. 1965. [Photo courtesy of Janine Dunn]

discovery that finally freed him from market gardening might never have happened had it not been for something he saw at the city market.

Double petunias were not common in the years before the outbreak of the Second World War, but they did make their appearance from time to time in Edmonton, grown from seed supplied by the Japanese. Simonet is reputed to have first seen them at the market when they were brought by a fellow vendor one Saturday to sell. Sensing both an opportunity and a challenge, Simonet applied himself to discovering how to produce seed that would result in 100 per cent double flowers. Dunn recalled the determination with which her uncle carried out his experiments. It would not have occurred to him, she said, to stop trying and he reached his goal in stages, not with one single discovery. In 1937, approximately three years after he set to work on the problem, he succeeded, a timing that ensured his reputation as a plant breeder and worked very well to his financial advantage. By the time war broke out in 1939, cutting off all trade with the Japanese, Simonet was in full production mode and ready to sell the much-coveted seed. In January 1946, when the George G. Ball company in the United States put out its seed catalogue, it was Simonet's "American 100% semi double flowering Petunia" that was featured on the front and back covers.[62] According to fellow plant breeder Fred Fellner, Simonet produced petunias in nine colours and went on to produce them in large-flowered double forms.[63]

The job of producing the seed, according to Simonet's niece, was tedious, time-consuming, and best done by women whose smaller hands gave them an advantage. Simonet's sister Marguerite was the first person to work with him in the seed production end of the business, but he also employed five young women to do the delicate work. Thousands of single-flowering plants were cross-pollinated by hand with double-flowering plants. Harvesting the seed at the end of the cycle was difficult work, said Dunn, but the return per ounce was high. Thus, while Simonet continued to work on new varieties of double petunias, he became the sole producer of seed that was sold all over North America and Western Europe at prices of up to five hundred dollars per ounce.[64]

Lilies, varieties of which are native to Alberta, were another of Simonet's many plant-breeding interests. In the early 1980s, when the Alberta Horticultural Advisory Committee was collecting plant material developed by Simonet for a planned collection at the Devonian Botanic Garden, his lilies were sought both for their historical interest and for their potential for "future plant breeding work."[65] Fellner knew more than most the potential for future plant breeding to be found in Simonet's lily collection. In an article first published by the North American Lily Society in 1979, Fellner told the story of Simonet's influence on his own development as a breeder, an influence that came partly through the transmission of ideas and information and partly through the transmission of actual plant material.

Fellner corresponded with Simonet during the 1960s about their mutual interest in lily breeding, learning something about the latter's breeding program. According to Fellner, Simonet's interest in lilies was first aroused by the native species *Lilium philadelphicum*. From Edmonton schoolteacher and lily breeder Fred Tarlton, Simonet received his first trumpet lilies, while his first Asiatic lily seed came from Saskatchewan nurseryman and plant breeder Percy Wright. Simonet had developed two black lilies and an outfacing deep pink lily called 'Embarrassment' before Fellner, who farmed for a living, was finally able to make his first "pilgrimage" to the Sherwood Park acreage in 1972. Fellner left, as did other such horticultural pilgrims, with a reward, in this case a dozen Asiatic lily seedlings and an equal number of trumpet lily seedlings, material from which he later developed a down-facing lily with up to thirty-six purplish-pink flowers that he named 'Robert Simonet'. But much more was to come, and Fellner's debt to Simonet was eventually expressed through the series of lilies he developed and registered from material he had obtained from his mentor's breeding program.

In the spring of 1973, Simonet gave Fellner hundreds of bulbs. Then, in the fall of the same year, he dug out seventy plants from an area he planned to dig over and took them to Fellner's farm in Vermilion. According to Fellner, Simonet "had no way of knowing just how great his work was at that time, for he neither visited other lily breeders nor bought any new named lilies." However, it was from this stock that Fellner developed and registered a number of lilies that he called the "Rescued Series," one of which, the beautiful red 'Lily Simonet', was again named after his mentor.[66]

Simonet's reputation may not be based on his work with roses, but there is evidence that he, like many other Edmontonians, had a strong desire to see so-called tender roses bloom in Alberta, a desire that was translated into an extended breeding program. In 1959, an article in the *Edmonton Journal* reported that Simonet was working with tea roses,

'Lily Simonet'. [Photo courtesy of Shauna Willoughby]

cross-breeding them to create hybrids that would be hardy in the Edmonton area.[67] An article in the *Alberta Horticulturist* seven years later reported that Simonet was still "experimenting with hybrid tea and hardy rose species to develop a hybrid with the characteristics of both."[68] But Paul Olsen, who wrote an entire article on Simonet's rose-breeding program for *The Rosebank Letter*, estimated that Simonet actually began to work on roses in the 1940s and that they became, for a time at least, his major breeding interest. According to Olsen, only two of Simonet's roses, 'Simonet's Double Pink' and 'Red Dawn', were registered, both of these by the Manitoba nurseryman and plant breeder Frank Skinner. Another two, 'Dr. F.L. Skinner' and 'Pink Masquerade', found their way into the nursery trade. Although Simonet never reached his goal of producing hardy hybrid tea roses, his breeding program, according to Olsen, influenced the breeding program of Dr. Felicitas Svedja of the Central Experimental Farm in Ottawa when she was working on the Explorer rose series. This happened after a visit to Edmonton in 1964 by Dr. Svedja, who obtained some Simonet crosses that were used to develop the Explorer *Rosa kordesii* cultivars. In addition, according to Olsen, Simonet's 'Dr. F.L. Skinner' and 'Pink Masquerade' are both present in the pedigree of another of the Explorer roses, 'De Montarville'.[69]

Nothing, it seemed, could limit or confine Simonet's horticultural imagination. No sooner had he achieved 100 per cent reliability with his petunia seed than he turned his attention to hollyhocks, producing what a former Edmonton Horticultural Society president, Fred Hilliard, described as an outstanding display of double hollyhocks for exhibition at the society's 1952 flower and vegetable show.[70] Around the same time, taking his cue from the local interest in growing gladioli as it was represented by several members of the Edmonton Horticultural Society, Simonet turned his attention for a while to this stately flower, producing a few varieties, the best of which was 'Simonet's Buff'. In the vegetable line, his most well known introduction was the 'Alta-Sweet' turnip, a sweet and edible rutabaga that is still on the market. Sweet corn, garden peas, and parsnips were also part of Simonet's vegetable breeding program. Like many Albertans and Edmontonians, he turned his attention to fruit growing, experimenting with rhubarb, apricots, raspberries, strawberries, and cherries. He developed a sweet, red rhubarb known as 'Strawberry Rhubarb' and produced a host of apple crosses, none of which were registered but many of which found their way into the home gardens of friends and, on occasion, into nurseries. For example, when Edmontonian Ken Riske decided to turn his family farm into a nursery, he paid a visit to Simonet, who was then suffering from severe memory loss. He left with several apple seedlings from which he eventually selected two varieties. 'Millstream', which produces a red apple similar to a 'Goodland' apple, and 'Strathcona Gold', which produces a golden apple, became staple offerings at his Millcreek Nursery. Always keen to increase the functional qualities of woody ornamentals or to increase the ornamental qualities of fruit-bearing shrubs, Simonet tried some unlikely crosses. At one point in 1983, the Alberta Horticultural Advisory Committee had collected "75 woody plants representing various intergeneric hybrids of apple, pear, saskatoon and mountain ash," all of them developed by Simonet.[71]

Although he was a private person whose interests revolved primarily around his work in horticulture, Simonet's contacts with the horticultural community in Edmonton were many and varied. Like others of his generation, he played an active role in the Edmonton Horticultural Society, even taking a position on its board of directors in 1943. He was frequently named honorary president or honorary horticulturist of the society and acted as the patron of its annual horticultural show. When asked, he made himself available to judge shows and gave many talks to society members. George Shewchuk, who acted as program director for the society for a few years, said the only problem with having Simonet as a speaker was that he became so wrapped up in his subject that it was sometimes difficult to stop him. Roy and Bea Keeler, long-time members of the Edmonton Horticultural Society and keen gladioli growers, came to know both Simonet and his wife on an intimate basis, an intimacy that began with repeated pilgrimages to his acreage for seed and plants. In the early days of their friendship, Bea gained the impression that Simonet was single and began sending him samples of her fresh baking whenever her husband planned to visit. In later years, when the Simonets and the Keelers would meet to play bridge, Simonet's wife assured Bea that her freshly baked pies were both welcome and delicious. Shewchuk also spent hours at Simonet's acreage, always returning home with something that was in an experimental phase. He marvelled at Simonet's inventiveness, particularly when it resulted in things like delicious yellow chokecherries growing on purple foliaged trees, or a cross between a rutabaga and a kohlrabi that grew cleanly on a pedestal. Hugh Knowles, who joined the Faculty of Agriculture at the University of Alberta in 1948 as a landscape and woody ornamental specialist, came to know Simonet at both his King Edward Park and Sherwood Park acreages. Knowles was also impressed by the breadth of Simonet's knowledge and interests and he admired the plant breeder's willingness to participate in programs and on committees initiated by either the University of Alberta or the Alberta Department of Agriculture. And Fellner, who learned about lilies from several breeders, wrote that his "greatest debt [was] to Robert Simonet."[72]

Simonet's accomplishments did not go unrecognized. In 1960, he was the tenth recipient of the Stevenson Memorial Gold Medal, awarded by the Manitoba Horticultural Association on an occasional basis to recipients who are deemed particularly worthy. In 1967, Simonet was made an Honorary Life Member of the Western Canadian Society for Horticulture in recognition of his meritorious service to horticulture in western Canada. In 1974, he was made an Honorary Life Member of the Agriculture Institute of Canada and, in 1984, five years before his death, he was elected to the Alberta Agriculture Hall of Fame and described as "one of the world's leading plant breeders."[73]

Simonet died at Edmonton in May 1989. Had it not been for his early success with double petunias, he might not have achieved either the credibility or the financial security to pursue his many and various breeding interests. But if financial security had, happily, followed his first major horticultural discovery, financial gain was never his motive. According to Dunn, her uncle cared far less for money than he did about achieving some horticultural breakthrough, an assessment that was corroborated by an article in the *Alberta Horticulturist*, which claimed that "[l]ack of concern for financial remuneration...is one of the traits for which Mr. Simonet is best known. His experiments and plant breeding projects have always been motivated by curiosity and a love of all things that grow."[74] Although the crosses that went into Simonet's late work went unrecorded, a fact that deprived him of full recognition for what he accomplished, his passion for the enterprise of plant breeding was given an ongoing boost through the horticultural prizes he endowed through the University of Alberta's Faculty of Agriculture.[75]

In August 1967, H.T. (Harvey) Allen delivered an address to a meeting of the Alberta Horticultural Association. He marvelled at the contributions made to both the province of Alberta and to the science of horticulture by "individuals whose vocation was not horticulture" but who, "through their love of the beauties of nature, their tremendous powers of observation and the challenge that their surroundings afforded…have given us a plant heritage that we can ill afford to forget." Allen, who came to the Lacombe Experimental Farm in 1947 as a trained horticulturist, was referring specifically to plant breeders, but he might well have included in his praise the many other "amateurs" who helped to change the face of their city, province, and even their country "through their love of the beauties of nature." People like Ramsay, Pike, Bugnet, and Simonet came to Alberta and the Edmonton area to stay. And, as Carpenter wrote in the introduction to his book of essays *Writing Home*, "[p]eople who live in a place and learn to call it home are too vigilant about that place to allow it to pass out of their control."[76]

5 The Edmonton Horticultural Society
Working for the City Beautiful

There are few cities, if any in Western Canada that lend themselves so well to beautification as the City of Edmonton, where the natural richness of the soil, ample moisture (doing away with the need of irrigation), and the long summer days, the sun shining for 18 hours per day, all tend to help the man who for his own sake, and for the sake of the City, seeks to beautify his surroundings with trees, shrubs and flowers.

—"Edmonton the Beautiful," *Town Topics*, 1913

The Horticultural Society has been working quietly and efficiently with a view to making Edmonton a City Beautiful, by the planting of trees, laying out grounds, and rewarding the quality and perfection of garden products in the Bench Show and encouraging the love of gardening by promoting competition in garden contests.

—GLADYS REEVES

SPECTACULAR AS IS ITS DRAMATIC SITE on the curving banks of the North Saskatchewan River, Edmonton has rarely been called a beautiful city, a fact city beautification activist Gladys Reeves lamented some time around 1930 when she wrote, "We have a natural beauty in our ravines [and] river banks, given us as a gift from nature to improve or to mar... unfortunately we have done more to mar them than improve them."[1] At no time was this failure to match the built environment to the natural one more evident than in the years just following Edmonton's 1904 incorporation as a city. Industry dominated the North Saskatchewan River valley, streets were unpaved and boulevards practically nonexistent. Bustle, not beauty, was the order of the day. But for those who planned to stay in the fast-growing city, mere growth was not enough.

Signs that city beautification was about to become something of a civic priority began to appear early in the twentieth century. The opening of Ramsay's greenhouses and florist business in 1906 gave plants and flowers a new role in the city, making them a sign of sophisticated and genteel living. In 1907, when Montreal-based consultant Frederick G. Todd put forward a vision of river valley reclamation and parks creation, it was welcomed as an implicit challenge to the hodge-podge development that had taken place in Edmonton prior to that date, particularly in its river valley.[2] In late 1908 or early 1909, Edmonton's rival city of Strathcona, situated within view on the south side of the North Saskatchewan River, organized a horticultural society and began to advocate for beautification in that community. Edmonton gardeners soon followed suit.[3]

On November 17, 1909, a "representative gathering of ladies and gentlemen interested in horticulture" gathered in the board of trade rooms to form the Edmonton Horticultural Society, one of the principal aims of which was to be the beautification of Edmonton. Not surprisingly, perhaps, Walter Ramsay, who had championed the cause of beauty both as an individual and as an entrepreneur, was made the society's first president. To add weight and legitimacy to the city beautification agenda adopted by the society, the Honourable

George Bulyea, Lieutenant-Governor of Alberta, became its honorary patron, while the Honourable John A. McDougall, a former mayor of Edmonton who had just been elected to the Alberta Legislature, was named honorary president.[4]

The prominence of city beautification in the minds of those who took an active role in shaping some of Edmonton's first institutions owed much to the so-called City Beautiful movement, an urban planning movement that arose in the United States in the late nineteenth century and spread to Canada, Australia, and England, before capitulating under the economic pressures of the Great Depression. In 1909, when the City Beautiful movement was at its most influential point in the United States, Edmonton was a young city with little public infrastructure. The notion that landscaped public parks and private gardens could in themselves inspire a respect for beauty among citizens and improve the quality of urban life, a notion promulgated by exponents of the City Beautiful, appealed to many Edmontonians. When it made city beautification central to its mission, the Edmonton Horticultural Society took on the job of making Edmonton not just a more attractive place but a better place in which to live. During the first three decades of the society's operations, and until well into the 1940s, the term "City Beautiful" continued to be called up by society members when referring to their organization's goals and accomplishments. But what links City Beautiful to city beautification?

In his book *The City Beautiful Movement*, William Wilson describes City Beautiful as an attempt by Americans "to refashion their cities into beautiful, functional entities." The aesthetics of the movement, he wrote, were "expressed as beauty, order, system and harmony" and its ideals were realized primarily through urban design. However, Wilson wrote, at its core, City Beautiful was a cultural movement, an attempt to "persuade urban dwellers to become more imbued with civic patriotism and better disposed toward community needs."[5] And it was the broad, cultural aspects of City Beautiful that adapted themselves so well to the aims and programs of horticultural groups across North

America, where gardeners often became ambassadors for community spirit and gardens were celebrated as expressions of community pride.

Wilson suggests, based on the American experience, that City Beautiful was a short-lived but influential movement, reaching its "heyday" between the years of 1900 and 1910. It was, he says, "in the broadest sense...a political movement," demanding "a reorientation of public thought and action toward urban beauty." To be effective, it required a working together of elected bodies and their administrations with "enlightened" citizens. And, in the end, the movement went "beyond structure and process" to include "citizen agitation and activism on behalf of beautification."[6]

In Edmonton, however, where the horticultural society was a key exponent of the philosophy, City Beautiful ideas were most powerfully expressed and embodied through the 1920s and 1930s and continued to exert an influence for decades afterwards. However, as was the case in the United States, those who promoted the City Beautiful in Edmonton did so to reorient public attitudes towards the importance of making natural beauty, as in gardens and trees, integral to their conception of urban development. And, following the American model, the horticultural society undertook to agitate on behalf of, or at least to promote aggressively, city beautification. Publicity campaigns for the society's annual shows and competitions, for example, were cast as attempts to raise public consciousness around the importance of creating beauty in everyday life. Society leaders, including people like Reeves, W.J. (Bill) Cardy, George Harcourt, Ernest Stowe, and Alfred Pike, incorporated the City Beautiful into their personal philosophies and value systems, making them particularly powerful representatives of the cause.[7]

A name often associated with the City Beautiful movement is the American urban planner and landscape architect Frederick Law Olmsted. Although Olmsted was not an exponent of the movement himself, Wilson argues, and even disassociated himself from some of its progressivist notions, exponents of the City Beautiful appropriated many of his ideas and practices.[8] Olmsted's many renowned parks and boulevard systems, for example, inspired his followers to propose such systems for towns and cities across North America. Edmonton's river valley parks system and the complex of trails and roadways that complement it owe an indirect debt to Olmsted. First proposed in 1907 by Todd, who had apprenticed under Olmsted as a young man, and elaborated in 1912 by the Minneapolis-based firm Morell & Nichols, also an exponent of Olmstedian ideas, Edmonton's river valley parks system has become a defining feature of the city and remains a work-in-progress.[9] Olmsted's intellectual legacy, on the other hand, including his belief that beauty exercised a positive influence on human thought and behaviour, his insistence on the "inseparability of beauty and utility," and his emphasis on the park as a "meeting and greeting area" or even a "site of class reconciliation," influenced the views and actions of key members of the Edmonton Horticultural Society, individuals whose primary motivation was a desire to make horticulture an agent of social change.[10]

In the United States, Wilson contends, the City Beautiful movement was rather quickly done in by a countermovement, a group of architects and planners who scorned City Beautiful's emphasis on aesthetics and preference for classical building styles. Wilson refers to this countermovement as "city practical" because its proponents argued for a more scientific, functional, and professionalized approach to planning cities.[11] The city practical approach triumphed in Edmonton as well and the classically imposing civic centre proposed by Morell & Nichols was never built. Nevertheless, City Beautiful ideas retained their relevance in the horticultural life of the city, championed by the horticultural society and upheld, within limits, by the city's elected officials and its administrations.

That in Canada the City Beautiful movement found its most accommodating home in horticultural societies around the country was recognized by Edwinna von Baeyer in an article written for the *Canadian Encyclopedia*. Von Baeyer notes that while the City Beautiful movement influenced various civic centre plans around the country and stimulated the creation of parks, parkway systems, and tree-lined boulevards, in its Canadian version it lacked both "an integrated philosophy" and "an articulate national spokesperson." Nevertheless, she concludes, the amateur side of the movement, led by citizens, was lively and active in Canada, working mainly through groups such as local horticultural societies. "These smaller groups," she wrote, "often effected greater change than the professionals, bringing to pass flower boxes on Main Street, street tree plantings, landscaping of public buildings, railway station gardens, allotment gardens and park creation."[12] And, indeed, this was primarily the front on which the Edmonton Horticultural Society fought, one tree and one flower at a time.

The horticultural society's profile as a long-standing community institution has been shaped by its dedication to the idea that private gardens, landscaped public areas, and tree-shaded boulevards go a long way (or perhaps all the way) towards creating either a City Beautiful or a beautiful city. If City Beautiful was the goal towards which horticultural society members aspired, individual acts of beautification were steps towards achieving that goal. Members of the organization modelled their belief that every garden planted is an act of citizenship and therefore a contribution to the overall well-being of the community. Successive boards of directors planted and maintained gardens in public places, took active roles in civic initiatives related to city beautification, and tailored their competitions to the never-ending drive to attract converts to their cause. Implicit in various iterations of the society's articles of incorporation, and reflected in the design of its programs, is the idea, upheld by Olmsted and by exponents of the City Beautiful movement, that beauty has an elevating effect on human behaviour.[13]

The fact that the horticultural society did not quickly devolve to become a mere gardening club, concerned primarily with the day-to-day interests of its member-gardeners, is due to four factors. First, ideals and values represented by the City Beautiful movement influenced the articulated mission of the society in 1909. In attempting to meet its original objectives, the society put itself on a path from which it has never significantly diverged. Second, and perhaps more important, has been the collaborative political context within which the society has functioned. Beginning a few years after its formation, the horticultural society was given offices in the Civic Block, Edmonton's de facto city hall at the time. Over the years it worked with the city on a number of major initiatives, some geared to the growing of food, as in the vacant lots program, and others geared to enhancing the city's appearance by planting trees, promoting spring cleanup campaigns, and, more recently, working to involve Edmonton in the Communities in Bloom movement. The society's outward focus owes much to this long-standing informal partnership with the city. Third, and not unimportant in maintaining its commitment to city beautification, has been the quality of the society's leadership. Leaders during the first half-century of the society's existence were energetic and effective proponents of the City Beautiful, who promoted gardening for its positive effects on the community. And, finally, without programs to convert philosophy into action, the society would not have been an active and influential participant in the ongoing project of city building through beautification.

The horticultural society's first articulated mission identified it immediately as an advocacy group for city beautification: "The objects of the Society shall be to encourage Horticulture, Arboriculture, Agriculture, Botany and branches thereof by lectures, demonstrations or other educational means, and to advise, recommend and encourage the beautifying of the homes of the city and the public institutions thereof."[14] So reads an account in the *Saturday News* on December 4, 1909, of the newly formed society's first board meeting and of the program it hoped to realize. For the society's first board of directors, the terms "education" and "beautification" appeared implicitly to be linked, with beautification being the desired goal and education the means of achieving it. While this formulation was retained in subsequent iterations of the society's mission, its reverse formulation, whereby beauty becomes the means and an educated population the desired end, has also carried weight with society members. Both formulations continue to inform the thinking and programming of the society.

In 1920, well after the Edmonton Horticultural Society's 1912 amalgamation with the Strathcona Horticultural Society, and two years after it joined with the Vacant Lots Garden Club to become the Edmonton Horticultural and Vacant Lots Garden Association, the society registered a new set of bylaws, one which reflected its expanded mandate while maintaining an emphasis on city beautification. Vacant lot gardening had been presented to Edmontonians in 1916 partly as a way to "develop civic pride and encourage beautification of the city."[15] The horticultural society retained this emphasis in the formulation of its new and expanded set of bylaws. "The cultivation of vacant lots for economic or beautification purposes" was listed in 1920 as one of the objects of the horticultural society alongside the "beautifying of homes of the City, the grounds of public institutions, and of the streets of the City." These objects were reiterated in the 1938 version of the bylaws, nicely reproduced in a mini-booklet form for members.

In 1973, under President John Barlow, a new set of bylaws was registered that suggested a slight divergence from the path of advocacy for city beautification. After more than fifty years during which it had officially operated as the Edmonton Horticultural and Vacant Lots Garden Association, the society reclaimed its original name, the Edmonton Horticultural Society. In addition to the name change, the objects of the society were altered slightly to reflect changing program emphases. Unchanged was the society's public service orientation; the 1973 bylaws included "the beautification of home grounds of public institutions, and of the streets of the city" as a primary objective. Downgraded as an influence on city beautification was the vacant lot gardening program, while the annual horticultural show and garden competition were recognized for their ability to reward the beautifiers, thereby stimulating an interest in gardening among the general public.

In 2004, when the society's bylaws were again rewritten, times had changed; society members feared that their club was being perceived by potential members as too seriously horticultural. Thought was given to removing the word "horticultural" from the society's name.[16] Although the name "Edmonton Horticultural Society" was retained, the society's mission statement was shortened and simplified, eliminating any references to city beautification. Gardening was no longer to be justified by the civic benefits it conferred but rather by the benefits it conferred on individuals, offering them a creative and healthful activity to occupy their leisure time. Nevertheless, although all references to city beautification were written out of the society's constitution, and although City Beautiful concepts had long since vanished from public discourse, initiatives designed to model and promote beautification through gardening continued to dominate the society's roster of activities. The long-standing tradition of working with the City of Edmonton to support beautification initiatives was retained and refreshed through new programs. In the end, the society's long commitment to gardening as a form of social action has prevailed over cultural pressures to treat it simply as a "lifestyle" option.

The horticultural society's historical relationship with the City of Edmonton goes a long way towards explaining its steady commitment to public service. It is a relationship based on a congruence of aims, both groups embracing the City Beautiful movement's emphasis on citizen participation as the key to success in bringing beauty, cleanliness, and order to urban life. What the city has taken from its association with the society is expertise, ideas, and volunteers to organize shows, hold competitions, administer a vacant lots program, and assist with tree planting and community cleanup campaigns. What the horticultural society has gained from its association with the city, on the other hand, has included recognition and status, a partner for some of its key programs, financial assistance, and, for many years, an address and offices in the middle of the downtown. It is odd, perhaps, that despite the symbiotic nature of this long-standing and largely informal association between the city and the society, only once was the idea put forward that the society should simply be taken over and absorbed into the city's administration.

This happened in February 1932, when Alderman Keillor asked the city commissioners how feasible it would be for the land department to simply absorb the horticultural society into its own operations.

After all, he reasoned, many of the properties included in the society's vacant lot gardening program belonged to the city, necessitating close co-operation between the society and city officials. The superintendent of the land department at the time was against it, arguing that the society could run the program more cheaply and efficiently by making use of volunteer labour, leaving the city free from expense and both morally and legally free to sell property under cultivation if it so wished. Also against the plan were Stowe, the horticultural society's president at the time, and Cardy, chair of the vacant lot gardening program. Stowe and Cardy were worried enough by Keillor's question to write Mayor Knott, explaining that any revenues taken in from the vacant lots program were immediately redirected to one or another of the society's city beautification projects, including the "transformation of the north east corner of the Market Square, from an unsightly weed-patch, to a beauty spot." The question of amalgamation was dropped.[17]

The society's credibility as an advocate for city beautification was reinforced and probably enhanced by its accommodation arrangements, particularly in the years during which it occupied office space in the Civic Block, the building erected in 1913 to house Edmonton's elected mayor and council, council chambers, and administration. There seems little doubt that, particularly before the completion of Edmonton's first City Hall in 1957, the society acquired prestige simply by virtue of its civic connections. A de facto city hall address reinforced the notion that the society's work and programs complemented and even extended city programs and services, a notion that was not far from the perceptions of succeeding society executives. The society styled itself as the horticultural conscience of the city, spreading the gospel of city beautification and encouraging citizens to take pride in their city's overall appearance.

Beginning in 1918, the year the horticultural society took over the vacant lot gardening program, the City of Edmonton provided the society with an office on the fifth floor of the Civic Block.[18] This co-location arrangement lasted until 1940, when the city architect and inspector of buildings took over the society's space to accommodate an expansion of his own department. For years afterwards, the city continued to provide the society with offices in one or another downtown building, including a return to the Civic Block in 1958, a year after the 1957 opening of Edmonton's first city hall.[19] But never was the working relationship between the two groups as close as it was during the twenty-two years of co-location. Between 1918 and 1940, the horticultural society not only ensured that weed control and a rudimentary form of city beautification were incorporated into the vacant lots program. It also collaborated with various city departments on initiatives such as a model garden in Market Square; a widely advertised spring cleanup campaign during the 1920s and 1930s; the creation of a tree-planting committee that included representation from the city, community leagues, and the horticultural society; and ways to incorporate gardening into relief measures during the years of the Great Depression.[20]

The horticultural society's efforts to bring Edmonton to prominence because of, not in spite of, its gardens, to make Edmonton a City Beautiful, have benefitted from ongoing, but conditional, financial support from the city. Important to the city has been the society's ability to represent all citizens on horticultural issues. In 1912, for instance, Edmonton's secretary-treasurer wrote to his counterpart in the society to announce that while the city's finance and assessment committee had authorized payment of a three-hundred-dollar grant to the society, the grant was contingent on the understanding "that [council and administration] only recognize one Horticultural Society throughout the entire city."[21] This point may have been made explicit because of the 1912 amalgamation of the cities of Strathcona and Edmonton, an amalgamation accompanied by the coming together of the two horticultural clubs that had sprung up on both sides of the river. But, in fact, the city has counted on the society to represent, in the most inclusive way possible, the horticultural interests of the community as a whole, and the city's financial arrangements with the society have been implicitly or explicitly contingent on this being the case.

Available evidence suggests that yearly operating grants made by the City of Edmonton to the horticultural society were always tied to the public service side of the society's operations. In 1926, for instance, the city paid the society fifty dollars over and above its regular appropriation to cover the "expenses of the Tree-planting Campaign."[22] Annual operating grants awarded by the city to the society are easily traceable in the City of Edmonton Financial Statements to 1925 and they show that the city tended to be more generous in its budget than it was in the actual awards, an indication of how precisely the society was made to account for any monies received.[23] Although yearly operating grants from the city never formed a significant portion of the society's budget, the fact that they were applied for year after year, rising to approximately eight hundred dollars in the late 1980s, can best be explained by the city's ongoing reliance on the society's contributions to city beautification. When Mayor Bury presented prizes to the winners of the society's 1929 annual show, he "spoke warmly of the work done by the association towards beautifying the city." He "hoped the spirit of beauty would spread throughout the city and cause Edmonton to be widely known for its gardens."[24] References to the society's contributions to city beautification were regularly incorporated into remarks made by mayors and councillors at society events and into advertisements placed by the city in the society's annual prize lists.

Even the vacant lots program, designed originally around the need to increase food supplies for export overseas during the First World War, was accounted for by the society largely in relation to the city's beautification agenda. A spreadsheet covering the years 1927 to 1931 suggests that what the city may have forfeited in vacant lot rentals was more than gained back through beautification projects undertaken by the society. While vacant lot receipts collected by the society during this period fluctuated between $2,600 and $3,100, program expenditures for the same period, all of which went to city beautification projects, fluctuated from $5,500 to $5,800. In addition to covering annual shows and competitions, the society's revenues went towards beautification and enrichment projects such as the distribution of free seed to new residents, the creation of a garden at an "Old Men's Home," and the purchase of trees for city planting.[25]

Small cash donations by the society to the city were not unusual during the years of co-location in the Civic Block. In 1938, for example, the society made a fifty-dollar donation to the City of Edmonton "towards the upkeep of the cenotaph."[26] In 1939, the year the King and Queen of England visited Edmonton, the society's president wrote to the mayor, offering a fifty-dollar donation towards a city beautification project, suggesting that it "be used for the planting of trees to commemorate the visit of Their Majesties in June." In 1940, the society made a fifty-dollar donation to the city "to be used in connection with City Beautification or any other object which the Commissioners may see fit to use it." And in 1942, another donation was made for the cenotaph.[27]

More often than not, donations made by the horticultural society to the City of Edmonton have taken the form of a particular beautification project. The first of these may have been the garden in the northeast corner of the old Market Square, located just south of what is now Sir Winston Churchill Square. Laid out in 1925 by society members, it matured and flourished under members' care. By 1928, it had become a "small beauty spot in the city centre." Over one hundred dollars was spent by the society to maintain it that year, a figure that probably represents an average yearly maintenance expenditure. Exactly how long the Market Square garden lasted is not certain, but, in 1932, President Stowe reported that it had become the model for other beautification projects.[28]

In March 1929, the horticultural society's secretary wrote the mayor and commissioners, asking permission to plant shrubs and perennials on the two islands in the crescent of Connaught Drive. Mayor Bury wrote back with an enthusiastic acceptance of the offer: "I am sure your Society will adopt the most effective form of beautification. This kind of work is very very valuable to the City and deserves not merely the gratitude of the citizens generally but all the encouragement and support which can be given."[29]

In 1932, the city asked the society for help in planting five gardens and was royally obliged. Relying on its contacts with growers, the society managed to have seven thousand plants donated, all of which it planted in the selected locations. In June of that year, George Buchanan, chair of the society's public service committee, wrote to Mayor Daniel Knott to ask the city to water in the new plants. Knott responded immediately, writing first to the city engineer, A.W. Haddow, to arrange the watering and then to Buchanan, expressing thanks on behalf of the city for the public service rendered.[30]

In 1967, the horticultural society built a rose garden in Coronation Park in honour of Canada's centenary. Fundraising for the project began in 1965, when Mr. J. McAfee donated one hundred dollars towards the project in honour of his wife, Hilda, a keen gardener who, before her death in 1965, had won many prizes for her roses. Mrs. Gladys Muttart, also a keen gardener and married to businessman and philanthropist Merrill Muttart, contributed another one hundred dollars. By the spring of 1966, when the society published its prize list for the year, a total of three hundred dollars had been raised, enabling the society to negotiate with the city for a "suitable location." In the spring of 1967, "[horticultural society] members planted between five and six hundred bushes" and installed a "suitable plaque." The Centennial Rose Garden was added to and maintained for several years until, in 1984, it was removed during the building of the Space Sciences Centre, now the Telus World of Science.[31]

In 1978, as a one-year only, and therefore somewhat anomalous, beautification project to mark the city's hosting the Commonwealth Games, the society planted a flower bed across from the Kinsmen Fieldhouse in Kinsmen Park, using red, white, and blue petunias to create the official games logo. A few years later, in 1985, society members used a bequest from James McAfee to build a rose garden at a local nursing home, an idea it repeated in 1986 when another rose bed was planted at a different seniors' residence.[32]

In 1995, however, society President Patrick Brown worked with John Helder, principal of horticulture for the city, on a collaborative beautification project at the Muttart Conservatory, a project that resulted in a

> *Edmonton Horticultural Society volunteers Bea and Roy Keeler planting a rose garden at St. Michael's Long Term Care Centre, Edmonton,* EJ, *August 22, 1985.*

long-term commitment by the society. Helder's idea was to transform the conservatory's extensive grounds by inviting groups from the community to plant and maintain specialty beds there. While the four climate-controlled glass pyramids featured in the design of the conservatory were intended to house plants from around the world in indoor settings, the grounds were developed to showcase the possibilities of the local. A perennial garden, designed and planted by the society in 1995, has been maintained by it ever since. One or more annual beds, sometimes used to trial new introductions and sometimes set out as show beds, were planted by the society each spring for more than a decade, maintained throughout the summer, and taken down in the fall. Somewhere along the line, two rose beds were added. The collaboration between the city and the society on the Muttart grounds project, formalized through the annual renewal of a signed Partners-in-Parks agreement, became the blueprint for an even more ambitious project, a four-bed strolling garden located on a busy path in the valley of the North Saskatchewan River.[33]

The Edmonton Horticultural Society Centennial Garden in Henrietta Muir Edwards Park could not have been realized as a project had it not been for the commitment of both the society and the city to making city beautification a community value. Planning for it began in 2005, the year the society began to negotiate with the city for a suitable piece of property, and it continued with input from both groups on matters of design, financing, timing, and ongoing maintenance. In June 2007, a month and year chosen so the garden could mature prior to centennial celebrations planned for 2009, a group of society volunteers gathered to plant the four beds, each with a distinctive colour theme. Upfront costs for the society included a twenty-thousand-dollar cash donation to the city towards the physical construction of the garden, funds it obtained from Alberta gaming revenues, and ten thousand dollars to buy the shrubs and perennials planted there. As per a Partners-in-Parks agreement, which is renewed annually and which spells out the obligations of both participating parties, the society maintains the garden, replaces plants as required, and sends out volunteers to weed, prune, and

Edmonton Horticultural Society Centennial Garden, Summer 2008.
[Photo courtesy of Marion Serink]

deadhead, while the city undertakes to mow grass, remove garbage, and water as required.[34] The horticultural society's work on the grounds of the Muttart Conservatory and its ongoing commitment to maintain the centennial garden it planted in 2007 are the most recent in a long line of city beautification projects, most of which were designed to embellish a piece of city property.

However, gardens need gardeners and values must be upheld and modelled by community leaders if they are to remain cultural norms. While the majority of the Edmonton Horticultural Society's leaders throughout its more than one-hundred-year history have embraced city beautification as a value and as an ideal to be striven for, a few can be singled out either for their achievements in moving the beautification agenda forward and/or for their passionate advocacy. In the former category is Cardy, whose lack of public image was counterbalanced by his effectiveness in office and his unfaltering commitment to all causes horticultural. In the latter category, that of public figure and passionate advocate, is Reeves, an attractive, unconventional, and outspoken champion of the City Beautiful who worked tirelessly to inspire others with her vision of Edmonton as a gracious and beautiful city.

Gladys Reeves was born in Somerset, England, in 1890, the youngest in a family of six children. In 1904, the family came to Edmonton, where her father, William Paris Reeves, promptly re-engaged himself with his favourite hobby, gardening. Gladys was close to her father and they shared an interest in gardening. But while her father was the consummate practitioner, spending his summers in the garden and his winters planning for the following year, Gladys became the consummate advocate, promoting gardening as a form of citizen action, her desired goal being the City Beautiful.[35]

Reeves left school a year after her arrival in Canada and went to work for photographer Ernest Brown. By the early 1920s, when she first joined the board of the horticultural society, Reeves had founded her own photography business, the Art League, and settled into a lifelong liaison with Brown, a married man, older than she, but with whom she remained close until his death in early January 1951. Between 1920 and 1928, years that Brown spent mainly in Vegreville, Alberta, Reeves devoted much of her spare time to working with the horticultural society and giving speeches about the importance of gardening and tree planting. Better remembered close to a century later as a photographer and companion to Brown, her accomplishments as an advocate for the City Beautiful may constitute a more important legacy to the city.

Reeves found what she needed in the horticultural society to facilitate getting her message out to the greatest number of people. In 1921, after criticizing the lack of publicity that had accompanied the 1921 flower and vegetable show, she joined the society's board of directors to work on the advertising committee. Advertising and publicity remained in her hands for several years, but it was in 1924, the year she spent as the society's first female president, that she was able to advance the City Beautiful agenda to an extraordinary degree. In the society's 1925 prize list, Reeves reported to members the accomplishments of her year in office, a year during which she claimed the society had been "working quietly and efficiently with a view to making Edmonton a City Beautiful." In addition to offering seven free lectures on horticulture to the public, the society had "co-operated with other organizations, such as the Clean Up Campaign and the Community Leagues, in the work of planting trees on the boulevards, schools and public places, and…undertaken the laying out of gardens at the Riverview pavilion, [and] the Y.W.C.A." Floral displays had twice been placed in the Hotel Macdonald to welcome special visitors to the city. A permanent secretary had been engaged to assist with the more than thirteen hundred vacant lot rentals handled that year. The 1924 annual garden competition had attracted 125 entries and the annual flower and vegetable show, held for the first time that year at the Riverview Pavilion, drew 1,204 entries and over four thousand paying visitors. Newspaper reports of it were glowing, garnering Reeves praise from her fellow committee members and compliments from the Lieutenant-Governor, who officiated at the opening ceremonies. The society's annual show, which Reeves claimed was "amongst the few large shows in the dominion," achieved levels of participation under Reeves that were maintained and even increased throughout the 1920s and 1930s, the years during which she retained the role of publicity chair. And, finally, she managed to convert the organization's 1923 deficit to a healthy surplus. In her year as president, Reeves injected a level of dynamism into the society's operations that it had hitherto lacked. She also formed a close working relationship with men and women who were willing, as was she, to extend the reach of the society beyond the garden to plant trees on city boulevards and to run an annual city cleanup campaign.[36]

The profile Reeves gained as president of the horticultural society likely helped her build a reputation as a champion of the City Beautiful. Judging by the notes she made for her speeches, which now reside in the Provincial Archives of Alberta, she took every opportunity to promote city beautification. In talks she gave to the Beverly Horticultural

Association, the Association of Community Leagues, groups of school children, service clubs, and various unidentified groups, she promoted tree planting and encouraged members of the public to take an active role in beautifying their own surroundings. She frequently spoke about the importance of educating and inspiring children in the arts of gardening and tree planting because children "will only have the inspiration which we pass on to them."[37] She worked for several years during the 1920s with the Keomi Club of Edmonton to judge beautification projects on school grounds. Although she was often critical of the results obtained on school grounds, she was far from critical of the Keomi Club; in notes for a speech, probably delivered between 1926 and 1928, she praised it for "desiring to add [its] weight to the activities of other organizations engaged in making Edmonton a City Beautiful."[38]

Of all Reeves's horticultural interests, tree planting was likely the one closest to her heart. She agreed with Oliver Wendell Holmes, whose writings she was known to quote, that "[w]e find our most soothing companionship in trees among which we have lived, some of which we ourselves may have planted." In undated notes for a speech, one of which was to be illustrated with slides borrowed from Professor Harcourt, she admitted that her only qualification as a speaker was that "I love Trees, I love beautiful surroundings; [and] I want Visitors to our City to take home with them the impression that the People of Edmonton must love their City or they would not have taken the trouble to make it lovely." She lamented the small amounts of money spent by the city on parks and boulevards and urged her audiences to put pressure on governments to do more.[39] In 1931, by which time most of Reeves's active work with the tree-planting committee had come to an end, a Montreal newspaper reported that in Edmonton that year children from thirty-five Edmonton schools had planted between 525 and 550 trees on Arbor Day, an initiative the reporter attributed to Reeves. "If there is a woman in Edmonton who should be able to predict the future of a tree it is

> Gladys Reeves, secretary, Edmonton Tree Planting Committee, c. 1925. [PAA B.7351]

Miss Reeves," wrote the reporter, "for busy professional woman that she is, she has given hours of her time, and boundless energy to the beautification of her city, with no other remuneration than the knowledge that the world will be a more gracious place, Edmonton a more beautiful city, because of her work." Reeves kept this article in her personal papers, a small indication of the satisfaction she took in being so judged.[40]

Although Reeves was secretary, not chairperson, of the small, but highly effective, Edmonton Tree Planting Committee, a group formed some time in the early 1920s to spearhead a tree-planting campaign across the city, she may have been its most influential member; she was certainly its most vocal one. This small group, made up of activists from the horticultural society and from various community leagues, worked with the city engineers department to encourage citizens to plant trees on their own properties and on public grounds such as boulevards, school yards, and public parks. They also obtained stock from local nurseries and sold it at wholesale prices at the city market. Tree-planting campaigns were integrated into the city's annual spring cleanup campaigns and were linked to the much-observed and publicly proclaimed "Arbor Day." In 1924, when Mayor Blatchford launched the city's first annual "Clean-Up Campaign," he highlighted its tree-planting component as the campaign's centrepiece. Ten thousand handbills were to be distributed to households, urging residents to clean up and beautify their properties and instructing them on the basics of tree planting. Approximately thirteen hundred shrubs and trees were purchased by the committee from Lacombe Nurseries and sold "at wholesale price" at the city market. The tree-planting committee supplied "some 700 young birch trees for the purpose of re-planting the City boulevards," explained a commissioners' report, trees that Reeves mentioned in a letter to Brown as having been donated by the Honourable P.E. Lessard, who "thinks our work is great stuff."[41] And, finally, the commissioners' report noted that a "special tree planting" was being planned by the committee for Arbor Day on May 7.[42]

Edmonton Tree Planting Committee, c. 1925. [PAA B.7331]

As soon as the 1924 cleanup campaign was announced, Reeves put herself at the centre of it. As secretary and chief spokesperson for both the cleanup committee and its tree-planting component, she maintained a close liaison with the city and with local newspapers. Her photography studio, located at 10348 Jasper Avenue, was listed on the cleanup committee's letterhead as "Offices of the Campaign."[43] The tree-planting committee, Reeves explained in notes for one of her talks, operated at first by digging native birches and evergreens from the woods and loading them into trucks loaned by "wholesale houses." "Supervisors for each street got every householder to come out and dig the holes on the boulevards; leaflets giving instructions how to plant were distributed." "[A]nd so," Reeves wrote, "with everybody working a vast work was accomplished." When the committee was criticized for planting native trees on boulevards instead of imported varieties, Reeves went on the offensive, claiming that native trees were not only cost-effective to plant but that "one of the prettiest Avenues in the City was an avenue of Birch Trees on 85th Ave."[44]

The "special tree-planting" for Arbor Day referred to in the 1924 commissioners' report was the subject of an article that appeared on the front page of the *Bulletin* on April 23 that year under the headline "Plan to plant trees at Memorial grounds." Members of the horticultural society, the article explained, were to meet that day with members of the "Great War Veterans Association" to discuss a ceremonial tree planting outside Memorial Hall that would "keep green the memories of those who fought for Canada."[45] The event, which involved planting a tree for each of the Edmonton battalions that took part in the First World War, took place on May 3, 1924, with Lieutenant-Governor R.G. Brett on hand to declare each tree "well and truly planted."[46] Reeves referred to this project years later in a speech urging citizens to involve themselves in similar projects. "When we planted the weeping birch round the Memorial Hall in memory of the Battalions who served overseas, it was hoped it would be the beginning of a Transcontinental Avenue or at least an Ave[nue] from here to Calgary." Her idea, one that is currently used by the city in a commemorative bench program, was that families wishing to commemorate a lost son or husband could donate a tree for the transcontinental avenue, a tree that would be identified first with a wooden board placed adjacent to it and ultimately with a plaque affixed to the tree.[47]

Realizing the limitations of native species as boulevard trees, and aware also that sources of native species were limited, the committee decided to expand its operations. In April 1925, Reeves wrote to the city on behalf of the tree planters to make two requests. The first was that the city, instead of providing the committee with trucks for hauling trees, provide it with a grant so that it could organize the transport itself. The second was for space that could be used for a tree nursery. The city commissioners, impressed with the fact that the committee had already been responsible for the planting of twelve thousand native trees, liked both suggestions. They recommended that council approve a two-hundred-dollar grant so that the tree-planting committee could arrange its own haulage and "that the city of Edmonton grant to the [committee] the use of the site set aside for a park, in the north section of the Hudson's Bay Reserve, for a period of ten years, for the beautifying of the City. At the end of said period the Committee to hand back the land in a good state of cultivation for park purposes, and surrounded by a double row of trees."[48] It was an approach sometimes criticized by nurserymen selling at the city market who claimed the city-owned nursery was unfair competition, but Reeves was fierce in her rebuttal; the horticultural society had stimulated an interest in planting trees, she claimed, which, "as soon as Edmontonians were fully awakened to the advantages of beautifying their home grounds," could only increase business for market vendors.[49]

Reeves's fertility of mind and ability to work with others appear to have been critical factors in many successful city beautification initiatives with which she was connected. In one of the many letters she wrote to Brown in 1926, she assessed her own contribution to the work of the horticultural society, concluding that "in the past seven years most of the constructive ideas [in the society] have come from me; and the others have worked more unselfishly because I did. It is something accomplished anyway."[50] She remained an active member of the society's board of directors for a decade, leaving it in 1932 but returning in 1935 to serve for another two years. Although 1936 marked the approximate end of her active work on behalf of city beautification, her "heart was with the movement," as she explained in a letter to the horticultural editor of the *Winnipeg Free Press* in 1933.[51] And for Reeves, as for many of her contemporaries in the horticultural society, to work for the City Beautiful in Edmonton was to participate in a movement, the goal of which was not simply to beautify the city but to convince the citizenry that public beautification was both a community value and a community responsibility. Moralistic and old-fashioned as she must have seemed in the 1970s, by which time a new generation of horticultural society leaders was in place, her contribution was recognized in her appointment as the society's honorary president, an appointment she held for four years prior to her death in April 1974.[52]

William James Cardy, who succeeded Reeves as president of the horticultural society and went on to serve in that capacity for twenty years, from 1925 through 1927 and from 1935 though 1951, never achieved his predecessor's public profile, but, like Reeves, he frequently employed the term "City Beautiful" to describe the ideal combination of city beautification through citizen engagement. And, while Reeves may have been City Beautiful's most vocal champion in Edmonton, Cardy was its most hard-working and dependable servant, from the time he joined the society's board of directors in 1924 until his death in April 1960.[53]

In the 1921 flower and vegetable show, a show which spilled out of its venue in the basement of the First Presbyterian Church on 105th Street and prefigured the move to more spacious quarters and larger, longer-lasting shows, Cardy's name was among the winners, not surprising for a "keen gardener" who often supplied perennials from his own garden to less experienced gardeners.[54] However, Cardy's talents and interests were better suited to organizing than to exhibiting and the 1921 win remained an anomaly. Remembered by Bea Keeler, whose husband Roy worked with Cardy to help found the Alberta Horticultural Association in 1951, as a small, quiet, white-haired man who always wore a business suit, Cardy, like Reeves, found in the horticultural society an organization through which he could work towards a better and more beautiful city. But where Reeves was an apologist and a proselytizer, Cardy was a behind-the-scenes worker whose meticulous attention to detail and reliable follow-through may have been responsible for maintaining the society's social relevance and its vitality through the difficult 1930s, the challenging 1940s, and the transitional 1950s.[55]

The themes that characterized Cardy's period at the helm of the Edmonton Horticultural Society emerged as soon as he began to take on executive responsibilities. First and most important, the society retained and even sharpened its focus on public service. Cardy worked well with others and, like Reeves, was able to create strong teams around the society's programs and special projects. A manager in the city's electric light department by day, Cardy understood the city's administrative structures and worked well with representatives from several departments. And, better than any other horticultural society leader, before or after his time, Cardy searched out and created links with other horticultural groups and agencies in the province. He developed a close working relationship with the University of Alberta's horticulturists, especially Dr. J.S. Shoemaker, collaborative relationships with several horticultural clubs in the province, and contacts within the provincial Department of Agriculture and both federal and provincial research stations. And, finally, Cardy worked to extend the horticultural society's reach into the community, making it a resource for gardeners and potential gardeners alike. The society's annual shows remained major events under Cardy's leadership, attracting large numbers of viewers from the general public. And, as chair of the society's vacant lots committee, a job he held even while he was the society's president, he worked closely with several city departments to ensure that lot rentals served a variety of civic priorities, only one of which was city beautification.[56]

The 1930s were difficult years for many families in Edmonton, but they were years during which Cardy was able to increase the horticultural society's relevance to the city and maintain an emphasis on city beautification. Increased relevance was achieved partly through programs designed to boost food supplies and to improve cultural practices, programs such as the vacant lots program, potato trials, and potato growing field days. But by working with the horticultural department at the University of Alberta on gladiolus and dahlia trials, contributing money to the university "to assist in establishing a test plot for Roses," maintaining the tradition of staging spectacular horticultural shows in August, holding and supporting specialty shows in a variety of venues, and producing and screening a movie featuring

Edmonton gardens, the society kept itself in the public spotlight. In August 1936, Cardy wrote to the mayor and commissioners to congratulate them on the "wonderful improvement made to public grounds within the last year or so, especially as represented by the grounds at the Power Plant and the Street Railway Barns." He viewed these improvements as an incentive to others.[57] That same year, he co-operated with the University of Alberta to open the trial grounds on campus to members of the public for a day, an idea that "was greatly appreciated by a large number of visitors." In his report on the 1937 operations of the society, Cardy expressed his thanks to the press "for the publicity given to further the work of making Edmonton a City Beautiful." And, anticipating a need for bodies to carry forward the society's work, he launched a membership drive, appealing "to the citizens of Edmonton to beautify our fair City." The colour movie of Edmonton gardens, which appears to have been made in 1937, added to in 1938, and shown several times in the city and elsewhere in Alberta, has unfortunately not survived as a relic of those years. In 1939, however, a very successful screening of this "pictorial record of Edmonton flower gardens" took place at the Capitol Theatre, an event which drew capacity crowds and from which several hundred people had to be turned away. And, finally, although the 1935 August show was reluctantly cancelled on account of a very early killing frost, the only such cancellation in the society's history, by 1938 no less than twelve exhibitions and competitive displays were held during the season, all, Cardy wrote in his report for that year, either sponsored or supported by "your society."[58]

The 1940s were challenging years for the horticultural society, which had to cope with declining numbers of lots made available by the city for gardening purposes, even as demand for them increased, a demand created partly by federal and provincial government food policy. Cardy's talent for collaborative organizational work was fully exercised as he worked first with the city on a rationing system for vacant lots and then with the provincial government's Department of Agriculture on a plan to promote growing vegetables. In March 1943, he organized a series of meetings throughout the city on how to grow a "victory garden." In November of the same year, the mayor referred to him a request from the chairman of the Agricultural Supplies Board in Ottawa for an assessment of Edmonton's victory garden program. The horticultural society responded that it had rented approximately three hundred and eighty acres of land through its vacant lots program in 1943 and the amounts of vegetables raised included an estimated seventy-five thousand bushels of potatoes and fifty-six thousand pounds each of cabbage and turnips. Carrots, parsnips, and beets were also grown in large amounts and Cardy did not attempt to put figures to the amounts of salad greens, tomatoes, peas, and beans.[59] And, despite the emphasis on vegetable production throughout the 1940s, the society's annual horticultural show continued to be the focal point of the calendar year. An increasingly collaborative arrangement between the Edmonton Exhibition Association and the horticultural society led to a change of venue, with the society contributing to the exhibition association's horticultural show in July and the exhibition association providing a venue for the society's annual show in August. In 1941, Cardy signed a memorandum of agreement with the City of Edmonton that would allow the society to construct a dedicated storage facility for its show props at the exhibition grounds, a facility that the society would have for its exclusive use, although, upon completion, it would revert to city ownership.[60] For Cardy, who worked to raise the profile of gardens and gardening in the social and political life of the city, the 1940s was an important decade, and one that led him to his last major effort on behalf of horticulture, the creation of a provincial organization that could act on behalf of all gardening clubs in the province.

The formation of a provincial horticultural association in 1951 broadened the context within which the Edmonton Horticultural Society was to deliver its message of city beautification and Cardy, who welcomed this change, was one of the principal architects of the new organization. In August and again in November 1950, Cardy attended meetings in Calgary to discuss the formation of a provincial association, meetings attended by representatives from other horticultural societies and, more significantly, by representatives from the provincial government Department of Agriculture. As chair of the new organization's constitutional committee, he proposed city beautification as one of its principal objects. In November 1951, the Alberta Horticultural Association was incorporated under the Alberta Societies Act and Cardy became its first president, a position he held for two years.[61] The formation of a provincial association, long desired by Cardy and many of his colleagues within the society, shifted the focus of the Edmonton Horticultural Society beyond city boundaries to take in the entire province.

Cardy died in April 1960, aged eighty-one, having devoted much of his working life and all of his retirement years to making horticulture an agent for the improvement of his home city and home province. By the time of his death, many new players had entered the horticultural scene, making city beautification not simply a matter between the city and its citizens but rather a matter of provincial interest. Horticultural research flourished at the University of Alberta and at both federal and provincial research stations around the province. Horticultural extension was being offered to home gardeners in the form of government publications, grants to organizations, and staff resources to home gardeners. Cardy lived to witness a period of government commitment to beautification and advancement through horticulture that he had helped to bring about and that endured for thirty years after his death before falling victim, in the early 1990s, to policy changes and funding cutbacks.[62]

The decades following Cardy's death saw a gradual, but permanent, change in the relative importance of some of the horticultural society's main programs, a change due at least in part to the increasing availability of horticultural knowledge and professional expertise. The annual flower and vegetable show, formerly a high point of the horticultural year for both exhibitors and for members of the general public who thronged to see it, gradually lost both its educational value and its appeal as spectacle until, reduced to a pale imitation of its former self, it disappeared altogether from the society's annual calendar in the late 1990s. In 1927, two thousand people had attended the Tuesday night opening of what the *Journal* called the "best flower show in Edmonton's history," a number that had swelled by another five thousand by the end of the three-day show.[63] In 1928, the same newspaper had boasted that Edmonton's annual flower show was ranked "among the finest on this continent" and ventured the suggestion that "every person with a sense of civic pride must thrill with pleasure at the city's flower-growing possibilities."[64] By contrast, the shows staged at the Royal Glenora Club in the 1990s were hard-pressed to garner exhibitors, spectators, and media coverage. The vacant lots program, too, fell victim to changing times. The thousands of vacant properties rented to Edmontonians every year from 1918 until after the end of the Second World War in 1945 were whittled down over time to a mere ten properties in 1989. The cost in time and money to administer the program outweighed the meager returns and the society decided to discontinue it.[65]

On the positive side, these were decades during which a provincial network of gardeners and horticulturists came together around the activities of the Alberta Horticultural Association. In 1949, the provincial Department of Agriculture appointed P.D. (Pete) McCalla as supervisor of horticulture, whose job included liaising with horticultural organizations across the province, facilitating access to grant money, and supporting horticultural initiatives originating in the voluntary sector.[66] McCalla, like Cardy, was present at the founding meetings of the Alberta Horticultural Association in 1950 and 1951 and remained a stalwart supporter of the provincial club throughout his career, acting as its secretary and newsletter editor for many years. The University of Alberta Devonian Botanic Garden, which became a reality in 1959 when a suitable tract of land was donated to the University of Alberta for the purpose, began to take shape and was opened to the public in 1967, providing yet another resource for the horticultural community. A Horticultural Advisory Board, appointed by the Minister of Agriculture in 1956, worked through its committees to ensure that the interests of urban gardeners were reflected in government policy and programs.[67] The provincial research station at Brooks, Alberta, included vegetables, fruits, and ornamentals for the home garden in its research program. Horticultural guides, which listed recommended hardy varieties for home gardeners, were updated regularly and made available to the public by the Department of Agriculture through the Faculty of Extension at the University of Alberta. "The Alberta Department of Agriculture stands ready to assist horticulturists of the Province with their work," read an advertisement placed by the Department of Agriculture in the Edmonton Horticultural Society's annual prize list in 1951, a message it repeated in subsequent years.[68] To fulfill its mission of city beautification, the Edmonton Horticultural Society strengthened its role as a mediator of knowledge by emphasizing its speaker program, continuing its long-standing practice of carrying out a variety of public and benevolent beautification projects, and maintaining links with all the components of this new, province-wide network of resources. When, in the early 1990s, the provincial government reduced its support for noncommercial horticulture and cut off its annual grants to the Alberta Horticultural Association, thus placing the gardening public in the hands of the private sector, the Edmonton Horticultural Society was again forced to reinvent itself if it was to maintain its prominent role as a proponent of city beautification.[69] Although the 1995 launch of a national organization called Communities in Bloom may not have seemed significant at the time, within a few years it became clear that it offered the Edmonton Horticultural Society a new and even wider context within which to carry out its original mandate of city beautification.

In many ways, Canada's Communities in Bloom program has provided the integrated philosophy and leadership that von Baeyer noted was absent from the Canadian version of the City Beautiful movement. Modelled on programs in countries like France and Britain, Communities in Bloom was launched in Canada in 1995 as a nonprofit organization "committed to fostering civic pride, environmental responsibility and beautification through community involvement." To accomplish its mission, the organization sponsors an annual competition that municipalities are encouraged to enter. Trained volunteer judges are dispatched by the national body to measure a given community's performance against criteria that include tidiness, environmental action, heritage conservation, urban forestry, landscape, turf and ground covers, floral displays, and community involvement. Municipalities may decide to compete at either the national or provincial level, depending on their competitiveness. But the end result is that all participating communities, regardless of how they place in the competition, receive feedback on their strengths and their shortcomings, feedback that helps to reinforce community spirit and boost community pride.[70]

Communities in Bloom is hardly the same thing as City Beautiful, but both have been useful to the horticultural society as banners under which to promote city beautification. While City Beautiful was the informing concept behind an approach to urban design, a concept that lent itself to appropriation by horticultural organizations in need of a slogan, Communities in Bloom began a century later as a carefully organized marketing plan, the success of which depended upon voluntary participation. City Beautiful was essentially led by the architects and planners hired by municipalities to create comprehensive urban plans; Communities in Bloom is a nonprofit organization whose goal is to raise awareness across the country about the value of making city beautification a civic priority. City Beautiful included an architectural focus that is entirely absent from Communities in Bloom. The two movements do resemble one another in their dependence on "buy-in" from the community at large. However, a century ago the horticultural society was able to take from City Beautiful what it needed to support its own city beautification initiatives. Today, the structure and organization of Communities in Bloom demands that municipalities take the lead, with voluntary and private groups taking on supporting roles. Whether the society's mission to promote city beautification has been made redundant by Communities in Bloom or whether Communities in Bloom offers the society opportunities to pursue that mission will always be open to debate by the society's leaders, but the partnership forged during the first decade and a half of Communities in Bloom gives some indication of the possibilities.

"In Bloom" sign, distributed to nominated yards by the Edmonton Horticultural Society, c. 2008. [Photo courtesy of Rodney M.B. Al]

Credit for ensuring that Edmonton was one of twenty-nine communities in Canada to participate in the first annual Communities in Bloom event in 1995 belongs to John Helder, principal of horticulture for the City of Edmonton at the time. But to make a case for Edmonton as a "Community in Bloom," Helder needed assistance and for that he turned to several groups, chief among them the Edmonton Horticultural Society. As the umbrella organization for all horticultural groups in the city, and as the organization with more ongoing projects and programs than any other, the society worked with Helder on Edmonton's first entry in the national competition and, within a few years, had tailored and modified its own programs towards the annual Communities in Bloom event.[71]

Each year, the horticultural society's president helps to host the national judges and tour them around the city. In 2005, 2006, and 2008, a horticultural society committee worked with Helder and others to organize an "In Bloom" show that gave organizations involved with beautification initiatives an opportunity to showcase their projects and accomplishments. Each year, members of the society volunteer their time to judge other municipalities, provincial or national, thus supporting the national program. Each year, the society maintains its own public gardening projects, including the Centennial Garden and the beds at the Muttart Conservatory, and ensures that they are in the best possible form for the judging. Programs run by the society for many years, especially its monthly speakers program, its annual garden competition, annual garden tour, twice yearly plant exchanges, and, from 2009 to 2012, its revival of the annual tradition of holding a flower and vegetable show in August, support the city's effort to gain recognition in the annual Communities in Bloom event. And, each year since 1999, the society has organized a Front Yards in Bloom event, another idea first proposed by Helder but one the society has taken on as its major contribution to the annual Communities in Bloom entry.[72]

Front Yards in Bloom, which recognizes gardeners who take the time to create gardens where passersby can appreciate them and honours those with the best gardens at a public awards ceremony, has replaced the annual flower and vegetable show as the horticultural society's most effective means of promoting city beautification. Launched in 1999 as a joint project between the City of Edmonton, the Edmonton Horticultural Society, and Canada Post, Front Yards in Bloom has expanded to involve a number of supporting organizations, including the Edmonton Federation of Community Leagues, the Edmonton Native Plant Group (formerly the Edmonton Naturalization Group), and Sustainable Food Edmonton. Front yards may be nominated by letter carriers, horticultural society volunteers, members of the general public, and community league volunteers. The horticultural society purchases and distributes a yellow "In Bloom" sign for each nominated garden, along with a letter of congratulation from the president of the horticultural society and an information package. The society co-ordinates several rounds of judging and arranges to have many of the best gardens photographed. And it organizes the final, or VIP, round of judging, which always includes the society's president, a city councillor, and a representative from Canada Post. With the support of local newspapers, particularly the *Edmonton Journal*, Front Yards in Bloom receives the media coverage that attended the August horticultural show in the 1920s, coverage that is essential, along with the open nomination process and the visible "In Bloom" signs, to spreading the desired message.

For the horticultural society, Front Yards in Bloom is a banner under which to promote community involvement in city beautification. The assumption underlying the program is that behind every front yard with so-called curb appeal is an engaged citizen, someone who, in Reeves's terms, demonstrates his or her "love" for the city by taking the trouble to make it "lovely." Questions of oversimplification aside, it is a formula that has informed the society's operations for more than a century, and, for Rodney Al, who wrote about the Front Yards in Bloom program in *A Century of Gardening in Edmonton*, a collection of short articles published by the society in its centennial year, Front Yards in Bloom "fully reflects [the society's] original commitment to community involvement and city beautification."[73]

On July 27, 2007, Michael Phair—former city councillor, long-time promoter of horticulture, and locally well-known activist on many social and political issues—addressed a celebratory gathering of members of the Edmonton Horticultural Society on the subject "Beauty and The Place." Phair referenced the Edmonton of 1909, then a city with a population of just over twenty thousand people. He suggested that although the passage of time brings with it superficial changes, underlying interests and values do not change; continuity from one time period to another can be traced beneath the surface. He wondered what difference gardening had made to Edmonton. For Phair, gardening indicates an ability "to engage with our surroundings." Gardens help to "distinguish the place where we live," whether this be home, neighbourhood, or city. They make the private public, an important gesture in a world inclined to present an anonymous face. For Phair, a beautiful place is an important indicator of the polity that lies beneath.

At the heart of both City Beautiful and Communities in Bloom lies the unarticulated assumption that beauty is not simply a superficial attribute, that behind a beautiful exterior is something we might think of as moral beauty. Simple, simplistic, or bearing the potential for both, this is the assumption that has lain behind one hundred years of work by committed leaders of the Edmonton Horticultural Society, working for the City Beautiful.

6

Waste Places
Vacant Lot Gardening in Edmonton

[I]t has been our aim to render useful or beautiful a vast majority of such city property as might otherwise have become an unsightly spot of weeds and neglect.

—W.A. STOWE, secretary, Edmonton Horticultural Society, 1935

*Go make a Garden for Victory,
It will keep you from getting jittery.
Take exercise on bended knee,
Weeding the carrots, the beets and the pea,
And help to reduce the scarcity.*

—*EDMONTON BULLETIN*, May 19, 1943

WHEN MRS. PLUGGER, the animated character in the wartime short-short film *He Plants for Victory* (1943) announces excitedly to her husband, seeds in hand, that she is going to plant a victory garden, he raises his head from the newspaper and rains hard on her patriotic parade. She knows nothing about gardening, he says. She does not even have the tools. Defiant, she proceeds with her plan. While Mrs. Plugger toils over her garden alone, her husband rallies the neighbours to a shared effort, a "Community Victory Garden," which soon blooms luxuriantly in contrast to his wife's miserable plot. At harvest time, Plugger takes his wife to the shared garden where she is suitably impressed with the results; they are depicted riding home on their bicycles with baskets full of produce. "A message about the importance of cooperation and knowledge sharing, especially during war time" reads a message from the National Film Board of Canada, which commissioned the one-minute, twenty-nine-second propaganda piece produced and animated by Philip Ragan. That is not entirely the way the film reads today. Today, we also see Mrs. Plugger as having been humiliated by the condescending and superior attitudes of her husband and we wonder why she is so cheerfully submissive.[1]

In 1943, the intended message of *He Plants for Victory* would have come through loud and clear to viewers, unclouded by feminist sensitivities to condescension that developed just a few decades later. Ragan did not need to explain his characters' eagerness to take up vegetable gardening. Urban food production was being promoted during the Second World War by all Western governments under the slogan "Plant a Victory Garden"; responding to the slogan was a way to demonstrate solidarity of purpose with the troops overseas while boosting the morale of those on the home front. We do not know if Mr. Plugger grew flowers as well as vegetables, or how he managed to motivate his neighbours, or why he waited to tell his wife about the community garden until the tomatoes were ready to harvest, or why, indeed, he did not involve her in the project. We do not know whether the land on which the community garden was planted was privately or publicly owned or whether it had previously been an ugly old vacant lot, a "waste place" that they transformed through their industry, although this latter scenario is implied. What we do know is that victory gardens were a political priority during the Second World War and that if this film, or a variation of it, were to have been made sixty years later, the message would have been different—"eating well on the 100-yard diet," perhaps, or "create a garden, create a community," or "good soil makes for a good life." Gardening, contrary to general opinion, is as political an act as it is personal.

Making "waste places...veritably blossom as the rose" has been an ongoing theme in Canadian urban history as it has in other parts of the world, including the United States of America.² Edmonton has been no exception. From the second decade of the twentieth century, when vacant lot gardening was first practised in Edmonton by the Vacant Lots Garden Club, to the second decade of the twenty-first century, by which time Sustainable Food Edmonton, Voices of the Soil, and other organizations had taken up the same cause and supplied it with a new set of cultural meanings, Edmontonians have been turning waste places into gardens.

In her book *City Bountiful: A Century of Community Gardening in America*, Laura J. Lawson emphasizes what she calls the "programmatic nature" of schemes to promote urban gardening. "While the idea of allotting land for gardening may seem straightforward," she writes, "in fact much organization and program development is necessary" in order to "coordinate gardeners, manage land, and facilitate educational or social activities." Although such programs can be run by groups, individuals, or government agencies, she suggests that any group administering an allotment program will need to rely "on a network of citywide, national, and even international sources for advisory, technical, financial, and political support."³

What Lawson concluded from her work on community gardening in the United States applies equally well in Canada. As fast as social, political, and environmental agendas for converting waste places to gardens appear, organizational structures emerge to convert these agendas into programs. Edmonton's vacant lot garden program, which flourished between 1916 and 1946 on the back of successive worldwide political crises, was sustained for an additional forty years by organizational structures originally put in place to deal with food shortages during the First World War.

George Harcourt, professor of horticulture, University of Alberta, n.d. [UAA 81-117-1-6]

According to the Ontario Horticultural Association, the vacant lot gardening movement in Canada began in that province in 1912, organized "[w]ith assistance from the Ploughmen's Association teamsters… to grow food for the needy."[4] In Alberta, vacant lot gardening began in 1914 when the Calgary City Planning Commission created a Vacant Lots Garden Club in that city, a club that endured until 1952 and that rented 3,299 vacant lots to its 2,366 members during the Second World War.[5] In March 1916, the Honourable Dr. Robert G. Brett, Lieutenant-Governor of Alberta, pitched the idea of a vacant lots club to Edmontonians in a speech he gave to a large group of "public-spirited citizens," including the mayor and representatives from many of the city's educational and interest groups, one of which was the Edmonton Horticultural Society.[6] Less than a week later, Edmonton had its own Vacant Lots Garden Club, directed by a board led by Professor George Harcourt.[7]

War was only part of the rationale supplied by the Lieutenant-Governor for the creation of the Vacant Lots Garden Club in Edmonton.[8] While the primary purpose of the new club was to convert unproductive city lots into productive gardens, its secondary purpose, to "develop civic pride and encourage beautification of the city," had already been claimed by the Edmonton Horticultural Society as its raison d'être. Overlap in the work done by the two organizations did not pass unnoticed.[9]

The Edmonton Vacant Lots Garden Club operated independently for two years, managing to allocate eight hundred lots in 1917. In 1918, however, by which time the Canadian government had become more insistent in its appeals to increase domestic food supplies, the vacant lots club expanded its organizational capacity by amalgamating with the horticultural society, a joining of forces that saw the society change its name to the cumbersome and rarely used "Edmonton Horticultural and Vacant Lots Garden Association." Harcourt, who supported the amalgamation on the grounds that it avoided duplication of work and stood to make a stronger organization, became the association's first vacant lots committee chair. His faith in the superior organizing capacity of the new entity appears to have been justified. During its first year as an amalgamated group, "2,818 lots were allocated to gardeners; 533 applications were received which could not be accommodated; and an additional 4,442 large blocks of land were rented to teamsters and others for pasture or to raise animal feed." It was an accomplishment brought to the attention of the Canada Food Board when its representative visited Edmonton in June 1918.[10]

Harcourt's commitment to vacant lot gardening, both as program and concept, could be regarded as a diversion from his principal work at the university, but it was not. For Harcourt, as for others of his time, the boundary between agriculture and horticulture was as porous as was that between the community and academia. Born in Ontario in 1864, educated at the Ontario Agricultural College in Guelph and at the University of Toronto, Harcourt first came to Alberta in 1905 to become Alberta's first Deputy Minister of Agriculture, a position he held for ten years before accepting an academic appointment at the University of Alberta. W.H. Alderman, in his book *The Development of Horticulture on the Northern Great Plains*, refers to Harcourt as "an important agricultural leader," one who was able "to develop enthusiasm in his students for the place of horticulture on the prairies and for the importance of gardening and home beautification."[11] Roger Vick, the first curator of the plant collection at the University of Alberta Devonian Botanic Garden, described Harcourt intriguingly as "a thrifty, quiet, considerate and modest man," who was also "a tireless entrepreneur, always enthusiastic, and occasionally rash."[12] Harcourt's interest in the development of hardy apple trees manifested itself shortly after his arrival when, in 1906, one of the ten branches of Alberta's Department of Agriculture created by his administration was devoted entirely to fruit experimentation.[13] Apple breeding was an interest he carried over to his experimental work at the University of Alberta and, although he was not entirely successful in his quest, his efforts were rewarded posthumously when, in 1955, one of his selections was named after him and found its way into the market.[14]

For Harcourt, taking on the job of organizing Edmonton's first vacant lot gardening program was both a response to a political priority and an opportunity to promote the benefits of home gardening. "Never in the history of the city has the real value of a small garden been so appreciated as this year," he wrote in his report to the horticultural society in 1918, and, almost as an afterthought, "it is safe to say that the product of the gardens of the city has done much to keep down the high cost of living and to allow greater stores of food products to accumulate for shipment overseas." Harcourt was gratified "to know that gardening in all its branches is coming more prominently forward for public attention and emphasis." He believed it would "result in a better appreciation on the part of the genial public of the great advantage a knowledge of these things has for the average person." Thus, while Harcourt's brief leadership of the vacant lots program was inspired by a political imperative, one that was clearly articulated in 1918 by the Canada Food Board, his interest in the program was based on overlapping agendas. Vacant lot gardens contributed directly to the war effort, but their longer-term benefit was the contribution they made to developing an enlightened citizenry.[15]

Although the Canada Food Board was not created until February 1918, just months away from the armistice that ended the First World War, its creation did not represent a change of policy for the government in Ottawa. As was explained in the introduction to the board's first report, presented in January 1919 at the end of the board's first year of operation, Canada's food policy during the First World War had already been developed by the office of the Food Controller; the creation of the Canada Food Board marked an endorsement and a continuation of that policy, which, simply stated in the opening sentence of the report, was "to supply the maximum of exportable foodstuffs to our Empire and the Allies during war." According to the report, "1918 was the year in which food as a war factor was proved to be only less mighty than were munitions." Food shortages in Britain and France threatened large numbers of people with levels of deprivation not easily imagined in North America. Rather than introduce food rationing, which the Canada Food Board judged would yield "infinitesimal" results, the policy was crafted to stimulate increased production of foodstuffs in Canada at the same time as conservation measures were put in place, "so that each month would see an addition to the exportable surplus." Vacant lot gardening was only one of the strategies put forward on the production side of the equation.[16]

For the Canada Food Board, as for Harcourt, the boundary between agriculture and horticulture was virtually nonexistent, a point of view reflected in the voluntary programs promoted and facilitated by the board. For instance, to boost food production on farms, especially where farmers and/or their sons were fighting overseas, the board initiated a campaign to increase the labour force by enlisting "Soldiers of the Soil," boys between the ages of fifteen and nineteen who were recruited to donate three months of their time during the summer to work on a Canadian farm. The organization of this program was shared between the food board and provincial governments, and, at the end of its first year of operation, the board reported that approximately twenty-five thousand boys had enlisted, almost all of whom had been satisfactorily placed on farms. The vacant lots and home garden campaign, designed to promote food production in urban areas, worked similarly. Government and community leaders across the country were asked to promote vacant lot and home gardening and, wherever possible, to put vacant lot gardening programs in place. Western Canada, where cities were newer and horticultural institutions assumed to be less well established, came in for special attention. The board arranged a "trip of inspection" to western Canada, with stops planned for "every city of importance between the Great Lakes and the Pacific Coast." The board's first report suggests that while it was impossible to pinpoint the precise monetary value of the produce from vacant lot gardens, it could roughly estimate production to have doubled from 1917 to 1918. Edmonton's own surge in vegetable production from vacant lots in 1918 more than contributed to that doubling.[17]

The organization put in place in 1918 by the horticultural society to manage its response to the national food policy yielded impressive early results. Although Harcourt's interest in vacant lot gardening was primarily to bring gardening "more prominently forward for public attention," the program was first presented to Edmontonians firmly within the context of national patriotism. "Conscript the Garden" ran an advertisement in the society's 1918 printed prize list.[18]

M.J. O'Farrell, a former real-estate agent whose passion for the vacant lots program matched Harcourt's, was engaged to assemble vacant land in the city and to get as much of it into production as possible. His first job, building the inventory, was accomplished with the help of the city assessor's office that provided access to contact information of property owners. At the end of that year, Harcourt summed up the work of the vacant lots committee, including staff member O'Farrell, as follows:

In all 7793 lots were passed through the office, practically 8000 lots. When one considers that on each lot the assessor's books had to be examined, the owner ascertained, written or phoned to, some idea of the magnitude of the task may be gained. In addition most of the lots had to be visited to be sure they were the right ones. This gives in the rough some conception of the very valuable service rendered to the community at large.[19]

It was a service acknowledged by the representative from the Canada Food Board, a Mr. Frederick Abraham, when he visited the city in June 1918.

Abraham's visit to Edmonton to tour vacant lot gardens, which took place before the gardens had fully matured, attracted attention in the newspapers, and his expressed views about the city's potential to become a major food production centre excited some. He "conservatively" estimated the value of Edmonton's vacant lot vegetable production at one hundred thousand dollars and saw "no reason why Edmonton should not, with urban labour, be the source of much of the vegetable supply for the province."[20] Abraham's visit, or at least the predictions arising from it, launched a reporter for the *Edmonton Bulletin*, one J.E. Pember, on a late July tour of vacant lot gardens, a tour inspired, at least in part, by the distinguished visitor's view that "many a more pretentious community fell far behind [Edmonton] in point of achievement." Pember appears to have been so overcome by witnessing both the quality and variety of produce grown in the city's vacant lot gardens that his inclination for hyperbole almost failed him. After a day touring the city, all he could see were "[g]ardens gardens everywhere until one loses count and the ability to distinguish between them and one's brain simply registers a composite impression of sturdy beets, carrots, beans, lettuce, turnips and rows and rows of potatoes neatly ridged up and due for a record yield of tubers which know no rival on the face of the earth." Pember saved one last rhetorical flourish to describe the transformation of a downtown lot by an unnamed "newspaperman" who had heard the "call" and felt "a desire for practical horticulture stir in his blood." The lot in question began as a "desolate and unsightly" heap of refuse from nearby building sites. Embedded in the construction debris were "[r]usted tin cans, broken crockery and bottles and rubbish of every description" and through it all grew "sturdy colonies of nettles and miscellaneous weeds which had obtained a foothold." Heroically, and Pember emphasized the heroic in his account, the downtown lot was transformed into "row on row of new peas already in the pod, beans which are well into blossom and potatoes which will not take a back seat from any in all Edmonton." There were "turnips just swelling into their white succulence, beets whose blood red globes invite the knife and fork and a lot of very thrifty cabbages and cauliflower," and more. For the *Bulletin* reporter, and presumably for others in 1918, downtown gardens were "keeping step with the thousands and thousands of other lots in Edmonton, which are engaged as truly in fighting the Hun as the soldier boy across the water, and are effectively backing up his efforts on the fighting line."[21]

Vacant lot gardening maintained its popularity in Edmonton throughout the interwar period, both as a tool in the hands of governments and among home gardeners, whose reasons for wanting extra gardening space were many and various. Government priorities were threefold. First, food shortages in Europe persisted into the 1920s, leading to an extension of the policies implemented by the Canada Food Board. Edmonton continued to do its part to encourage maximum food production through its vacant lot garden program. Second, the Great Depression of the 1930s brought with it a new set of social and cultural priorities. The City of Edmonton, the horticultural society's partner in the vacant lots program, identified vacant lot gardening as one means of delivering aid to the unemployed. Throughout most of the 1930s, the horticultural society's vacant lots committee worked with H.F. (Harold) McKee, manager of the city's special relief department, to tailor the vacant lots program to the needs of Edmontonians on relief. And, finally, throughout the entire interwar period there was the so-called weed menace to contend with. To retain the privilege of running the vacant lots program, the horticultural society was obliged by the city to ensure that all properties it administered were kept free of noxious weeds.

In the years immediately following the end of the Great War, O'Farrell ensured that the horticultural society's vacant lots program maintained its emphasis on food production. In 1920, he sent a form letter to property owners in the city, asking their permission to use properties previously donated to the program and encouraging them to "kindly add additional lots that you may care to allow to be used for food production purposes this year." He sent the same message to property owners who resided out of province. In June 1920, for instance, he sent a letter to a Mr. Arthur Buswell in Southsea, England, to request the use of the latter's Edmonton lot for vacant lot gardening. "Since the war has ceased," he wrote, "the world...finds itself in a serious predicament in the nature of a world-wide food shortage." He continued:

> In some countries in Europe fabulous prices are being charged for food and great distress is resulting, even starvation. In our own country prices are much inflated and there does not seem to be any immediate prospect of a decline. For these reasons it is as patriotic a duty to make the soil produce its maximum supply as it was during the days of the war. The Government of our country is urging the people to produce and conserve and in harmony with this advice the Edmonton Horticultural and Vacant Lots Garden Association feels that it still owes it to the community to assist citizens in every way possible to grow as much food as they can for their own use and for the use of others.[22]

O'Farrell remained in charge of the vacant lots program until 1922, after which the emphasis placed by the society on food production in the national interest waned gradually, to be replaced within a decade by new threats posed by the Great Depression.

The famous stock market crash of 1929 launched a decade of hardship for individuals and for the municipalities in which they lived. As revenue for municipal governments such as Edmonton's declined, numbers entering the ranks of the unemployed grew, putting a strain on the city's finances. To cope, Edmonton established a special relief department in 1932 to implement a variety of emergency relief measures, one of which was the so-called relief gardens. Vegetable gardens were widely recognized during the Depression as a way to boost a family's food supply, and McKee, head of special relief, adopted relief gardens as one of several measures intended both to improve the lot of those on relief and to reduce the financial burden carried by the city. Not only did McKee work with the horticultural society on adaptations to its vacant lots program, he also tried a larger-scale gardening project with a view to stockpiling vegetables for winter distribution to families in need. In May 1933, for example, the *Journal* carried an article about thirty-five acres of land the relief department planned to have in cultivation "to grow stocks of staple winter vegetables which will be used for purposes of direct relief next winter." The gardens, divided among five communities, including fifteen acres in Belgravia, were to be cultivated by "work parties of men on relief" under the supervision of Thomas Housley, the city weed inspector. This particular initiative, the newspaper reported, was "quite apart from scores of vacant lots granted to families on relief for the production of fresh summer vegetables and winter supplies," a reference to the program the horticultural society began to adapt to the needs of those on relief as early as 1930, two years before McKee's appointment.[23]

Delivering the relief garden program to significant numbers of often reluctant and inexperienced gardeners resulted in an increased workload for the horticultural society's vacant lots committee and a concomitant decline in the society's annual revenues, small hardships accepted by the society with barely a mention in the annual program reports it sent to the city. Beginning in 1930, when the society first began providing lots free of charge to those who could not afford to pay, and ending in the early 1940s, when the American military presence in Edmonton resulted in decreased numbers of lots being made available to the society to let out, the vacant lots program was modified to suit the needs of relief recipients. While numbers of gardens allocated free of charge in 1930 rated a mere mention in the society's correspondence with city officials, the number in 1931 had risen to 350 and continued to rise until 1936 when over a thousand lots, more than a quarter of the society's inventory, were given out free of charge. In 1939, the number of lots supplied to those on relief had declined slightly to 962, as compared to just over thirty-three hundred lots rented out in the usual fashion. By 1941, however, only 494 lots were given out on a relief basis and a new priority, the victory garden campaign, began to take precedence.[24]

McKee administered relief programs with a rigour that provoked criticism from many quarters, including some council members. Praised in 1935 by the equally abrasive Mayor Clarke, who defended McKee in council and recommended him for a pay raise at a time when most city employees were receiving wage cuts of up to 16 per cent, McKee often found himself at the centre of controversy on account of his apparent assumption that most relief recipients were lazy.[25] He was not popular among the ranks of the unemployed, who were often humiliated by policies approved by council and administered by the special relief department. Perhaps the least controversial of all the relief programs was the vacant lots program, although, in retrospect, its delivery reinforced the very stereotype of the relief recipient projected by McKee. Relief gardeners were often portrayed in correspondence from the horticultural society as having been ignorant of good gardening practice and indifferent to acquiring the skills.[26]

Throughout the 1930s, members of the horticultural society's vacant lots committee entered fully into McKee's muscular approach to relief delivery. In 1933, for instance, by which time 801 gardens were given out to relief recipients, the society introduced special classes to its annual garden competition that were open only to relief gardeners. The rather dismal participation numbers were communicated to McKee in a letter sent by the society:

> *We regret very much the lack of interest taken in these competitions. The prize money was donated by members of our society, and with over eight hundred lots granted to people on relief through us, we got seven entries in the vegetable garden and five in the potato lot. You can see with us that this is very disappointing and gives us very little encouragement to continue this competition another year.*[27]

Nevertheless, despite the disappointing levels of participation in 1933, the annual competition was continued, and renewed efforts were made to entice relief gardeners to enter. The city was divided into four quadrants with four sets of winners, an arrangement that multiplied the prizes and allowed results to be compared from one quadrant of the city to another. More classes were introduced, including a class for flower gardens. In 1937, council put up fifty dollars in prize money and, in May of that year, Commissioner Gibb, who characterized the competition as a direct response to council's expressed desire to encourage good gardening practice among the unemployed, wrote McKee to approve the arrangements.[28]

Unfortunately, council's investment in the horticultural education of the unemployed yielded unimpressive returns. Of the forty-eight registered competitors that year the society's judges noticed a "lack of proper cultural methods" in about 50 per cent of the gardens, with excessive weediness being one of the problems.[29] Although 1938 saw an encouraging rise in the number of competitors, the results were even more disappointing. Society judges complained that 50 per cent of the gardens "showed a distinct sign of lack of care," with another 25 per cent being only marginally better. Flower gardens were even worse than vegetable gardens, with almost half of them judged as "not worthy of scoring."[30]

Luckily, the annual competition cannot be taken as the sole indicator of the success or failure of the relief garden program. Correspondence from financially stressed gardeners and addressed to the city during the 1930s suggests that vacant lot gardens sometimes sustained families through periods of unemployment. On April 10, 1930, for example, a Norwood gardener, who had lost a lot he had gardened for eight years due to his inability to pay the small rental fee, wrote to the city. The lot in question, which had yielded "potatoes and truck to keep me in the winter when I was out of work rather than ask for relief," had been reallocated by the society to someone who could pay the fee. The

> *Antoinette Grenier, c. 1925, 10660–97th Street, Edmonton.*
[Photo courtesy of Antoinette Grenier]

following day, the gardener's wife wrote to the mayor both to reinforce her husband's story and to add further detail about their straitened circumstances. A few days later, Mayor Douglas wrote the couple to say that he had gone to the society and obtained a promise to restore the lot.[31]

Even for families on a more stable financial footing, vacant lot gardens often became an integral part of their family economy. Antoinette Grenier was six years old when, in 1925, she moved with her French-speaking parents to a small house on 97th Street, just south of 107th Avenue. Grenier remembers the area as having been ethnically and culturally mixed, a mix reflected in the range of produce grown in the backyard and vacant lot gardens in the area. Her own parents, she said, relied on the garden's contribution to the family income. To extend its productivity, they frequently applied to the horticultural society for a vacant lot located just across the back lane from their house. In addition to potatoes and carrots, they grew cauliflower, beans, peas, cabbage, and salad greens. Madame Grenier canned vegetables for the winter and the family luxuriated in fresh produce throughout the summer. From gooseberry and red currant bushes, which the Greniers had purchased from a nursery in Ontario shortly after they moved to 97th Street, the family harvested fruit for winter preserves. The work of the garden was shared among members of the family, with Antoinette learning at a young age how to weed the garden and how to sow and tend the plants. These basic skills, methodically exercised, made the Greniers model renters for the horticultural society in its ongoing struggle to ensure that the properties it administered were productively used and kept weed-free.[32]

The city's preoccupation with the "weed menace," a preoccupation shared by all levels of government, was acquired long before the vacant lot gardening movement appeared to assist with the urban manifestation of unwanted transplants. According to Clinton Evans, the first weed legislation in Canada was passed in Upper Canada in 1865, a response to the rapid proliferation of species imported, deliberately or accidentally, by settlers arriving in Canada from overseas.[33] Antiweed rhetoric developed in Ontario in the late nineteenth century and spread to the

West, where, according to Evans, agricultural practices favoured the proliferation of invasive species. On August 20, 1894, an article appeared in the *Bulletin* on one of these species, the Russian thistle, along with the information that the Northwest Legislative Assembly was about to bring in legislation to deal with noxious weeds.[34] In 1900, the Northwest government set up a tent at the Calgary Fair to exhibit specimens, a first step in what was planned as a "vigorous campaign of education on the subject of weeds."[35] A year later, in 1901, this educational initiative was followed up with the publication of the pamphlet "Noxious weeds and how to destroy them." Farmers were warned that the only way to control weeds was "by constant attention and by adopting methods in accordance with the nature and habits of growth."[36] City dwellers, too, many of whom were carrying on agricultural pursuits within the city limits, were subject to the same legislation and the same dire warnings about the dangers of allowing weeds to proliferate. As early as 1901, the weed inspector for the territories had "an assortment of noxious weeds on exhibition in the *Bulletin* window" for the edification of all city gardeners and farmers.[37] By 1908, a year before the Edmonton Horticultural Society came into being and seven years before the Vacant Lots Garden Club was formed, Alberta had a grand total of seventy-five weed inspectors at work on the problem.[38]

Not surprisingly, perhaps, ensuring that all vacant lots in its inventory were kept "entirely free of noxious weeds" was the primary condition imposed by the city upon the horticultural society, a condition that was sometimes as difficult to fulfill as it was rigorously enforced. In 1934, for instance, the society reported having particular problems controlling weeds on lots given out to relief recipients. That same year, in response to the problem, the weed inspector decided that no "odd" lots should be broken for new gardens: "If existing lots are not sufficient to meet the demand, I would recommend that a whole block be broken up and given out in lots. Then, if not required for garden lots it could be rented for grain or seeded to grass and so create revenue."[39] The weed inspector's recommendation may have been made in the spirit of administrative efficiency, but for the horticultural society, which dealt with thousands of gardeners, many of whom lived adjacent to or near "odd" lots, the recommendation could not have been an easy one to implement.

To retain control of the vacant lots program, the modest revenue from which rapidly came to play a critical role in the society's annual budget, the society engaged with the city in a yearly performance appraisal process that involved a good deal of to-ing and fro-ing on the weed issue. Renewal of the contract was not automatic and hoops had to be jumped through on both sides. In November 1926, for example, the society's president and its secretary wrote to the city commissioners requesting a renewal of the contract: "We have co-operated with the City weed inspector, who at our annual meeting gave us a talk on this subject with the result that it was decided to hand to each renter a printed notice calling attention to the importance of keeping the lots free from weeds." It was now up to the commissioners to verify the society's claim to have complied with all of the terms of the agreement. S.B. Ferris, superintendent of the land department in 1926, was asked for his opinion on the society's performance. He pronounced in the society's favour: "The Horticultural Society renders considerable assistance in keeping the cultivated lots in continued use for gardening purposes and preventing them from growing up in noxious weeds." Finally, the commissioners made their recommendation to council, which as soon as it had been approved, was communicated to the society in writing, with an inevitable reminder of the need to work closely with the weed inspector and to enforce all weed bylaws in the upcoming season.[40]

Weed control remained a dominant theme in all surviving correspondence between the city and the society, but just as it was temporarily displaced during the 1930s by the crisis of unemployment and the decision to adapt the vacant lots program to the needs of relief recipients, so it was displaced for a few years during and immediately after the Second World War when urban agriculture returned to the national political agenda as a way to boost food supplies depleted by war. "Victory Gardens," a term to parallel the equally catchy "Victory Bonds," and one that came into wide use in the 1940s to describe vegetable gardens planted for patriotic reasons, was actively promoted by governments at all levels. The National Film Board's 1943 release of *He Plants for Victory* helped to get the message out. However, it was left to organizations such as horticultural societies and to individuals such as the poetically challenged Edmonton resident who urged readers of the *Edmonton Bulletin* on May 19, 1943, to "[g]o make a Garden for Victory" to translate the victory garden campaign into increased urban food production.

Daily newspapers did their part to promote victory gardens. In April 1943, in an article headlined "Victory Food Garden Guide," the *Bulletin* provided practical advice to its readers on how to plant a victory garden and what to plant in it. Detailed planting instructions were provided for each recommended variety and, for those who had never planted a vegetable garden, a layout was recommended.[41] Perhaps it was this article that inspired a so-called hard-headed businessman to take up the hoe. In May 1943, W.T.H. sent the *Bulletin* a poem entitled "My First Victory Garden." It began:

My Garden is a bit of creation,
In which I get recreation.
I sow the seeds, remove the weeds,
And gather some information.[42]

Fortunately, for the paper's readers, this businessman/poet/gardener appears to have applied himself more seriously to the craft of gardening than to that of composing verses. Still, by the end of July, the *Bulletin* was able to report that W.T.H. was not the only Edmontonian "to reap the benefit of [his] foresight and industry." Vegetable crops in the city were said to have done well, including heat sensitive tomatoes that were "getting to a good size and g[a]ve promise of escape from the altitudinous price of imported supplies." "It is safe to say," concluded the writer of the article, "that no one who put in a Victory Garden and took care of it now thinks he wasted his time."[43] Hilda McAfee, an experienced gardener and member of the horticultural society whose affinity for the pages of the newspapers and gardening magazines was frequently indulged by her success in competitions, explained to a newspaper reporter in July 1944 that she had cut back the lawn on her relatively small city lot to incorporate a victory garden alongside the lily ponds, rock gardens, and a host of exotic plant varieties. Cunningly, according to the reporter, she had planted scarlet runner beans, both "to help out the victory garden" and to do double duty as a decorative feature.[44]

Responsibility for implementing the victory garden campaign in Edmonton fell primarily to the Edmonton Horticultural Society, which wrote the city commissioners in March 1943 to explain that it had been asked by both the "Dominion and Provincial Governments" to "assist them in encouraging the growing of vegetables as a war effo[r]t." Working closely with the provincial Department of Agriculture, the society planned a series of meetings at locations around the city to explain the program and encourage participation.[45] The pressure for relief gardens had all but disappeared, leaving the society free to focus its efforts on the war. In November 1943, the society responded to a request from A.M. Shaw, chairman of the Agricultural Supplies Board in Ottawa, for a report on the success of the campaign. According to the society, approximately 380 acres of land had been planted in gardens during the summer of 1943 through the vacant lots program, although it was "impossible to say" how many of these had been a direct response to the campaign. With more certainty, the society estimated impressive production figures for the vegetable crops, figures that did not include salad greens, beans, peas, or corn. In addition to seventy-five thousand bushels of potatoes, Edmonton's vacant lot gardens had produced thirty-seven thousand pounds of carrots, twenty-eight thousand pounds of beets, fifty-six thousand pounds of turnips, twenty-eight thousand pounds of parsnips, and fifty-six thousand pounds of cabbage. "We are of the opinion," the society's report concluded, "that any effort put forth, to encourage the public to produce at least part of their requirements of vegetables, is worthwhile and believe, the project had considerable benefit."[46]

The horticultural society's apparent accomplishments on the victory gardening front were somewhat surprising in view of its inability to expand its inventory of vacant lots. The influx of Americans to Edmonton during the war, many of whom were engaged with the northwest staging route to Alaska, and the remainder of whom came to build the Alaska Highway, had put enormous pressure on the city's inventory of disposable properties. In March 1943, society President Cardy wrote to the commissioners regarding the shortage of garden plots, a shortage he attributed to the fact that so many lots were "being taken over by the American Armed forces...or...sold." A wartime housing project planned for the Norwood area, for instance, was expected to drastically reduce the numbers of vacant lots available for rent in that gardening neighbourhood.[47] At the time, the encroachment on inventory may have seemed like a temporary setback, something that could easily have been remedied after the war and when the American influx had receded southwards, but the Second World War was followed by a new status quo that undermined the vacant lots program and precipitated its slow decline.

The discovery of oil at Leduc in 1947 is as good a marker as any of the changes that came to Edmonton after the Second World War. There was a new optimism about the future combined with a steady rise in Edmontonians' standard of living. The 1955 opening of Westmount Shoppers' Park, Edmonton's first enclosed mall, catered to changing shopping habits and a shifting preference for purchased over home-grown produce.[48] Changes in family structure and organization meant that women who took jobs outside the home had less time to tend gardens and preserve produce for winter consumption. The value of homegrown produce declined as out-of-season vegetables became available year-round. So, despite the continuation of the victory garden campaign after the war's end, with vegetable gardening being promoted as a bulwark against the rising cost of living, the Second World War proved to be the beginning of a long and drawn-out end for the society's vacant lots garden program. In 1947, the society reported having rented out only 1,590 lots, a high enough number in itself but a considerable decrease from any of the figures achieved between 1918 and 1945.[49] Lacking any political or moral content, the vacant lots program lost its popular appeal. While those who had lived through the Great Depression and the Second World War remained confirmed in their gardening habits, later generations saw no reason to produce for themselves what they could acquire so easily with less effort.

Ethel Dorin was one Edmonton gardener who noticed this change. Born in Edmonton in 1924 and raised at Chip Lake, Alberta, Dorin assumed that gardens were simply a way of life. Her father had been a keen gardener. Not only did her parents maintain a large kitchen garden, which made important contributions to the family's food supply, but her father managed to keep a decorative front yard flower garden in immaculate condition. Following suit, Dorin and her husband, Walter, always kept a vegetable garden at their home on what was then the eastern edge of the city in addition to the prize-winning gladioli and dahlias Walter grew for show. When Walter retired in 1982, after a career as sales manager for Pike Seeds, the Dorins bought a new house in a community in northeast Edmonton and Ethel was astonished to discover how few of her neighbours maintained vegetable gardens. She and Walter no longer considered themselves serious gardeners, but they kept a vegetable garden that they extended by renting from the city a tiny bit of land adjacent to their property. Every fall, Walter painstakingly picked and pitted the sour cherries from a tree he had obtained from Robert Simonet. He packed them in freezer bags for Ethel to use during the winter in cherry pies or cherry cheesecakes.[50]

Roy Keeler, president of the horticultural society for most of the years between 1957 and 1971, along with his wife, Bea, lived according to values similar to those of the Dorins. Roy and Bea would never have considered not having a vegetable garden, despite the vast amount of time they both expended on Roy's avocation as a gladiolus grower and prize winner at horticultural shows across the country.[51] And yet, if the Keelers and others of their generation were aware of the diminishing interest in vegetable gardening, even among horticultural society members, they were unable to generate a new and compelling agenda for the vacant lots program. In the absence of external forces to supply it with meaning, the program continued to wither away. In 1956, the society had only 350 lots to rent. By 1959, the number had shrunk to 215. In 1967, in an effort to generate more income from the program, the society raised the annual rental fee from two to three dollars. In 1979, a gardener had to pay twenty dollars for a vacant lot. By 1986, the number of lots in the society's inventory had shrunk to forty-eight. In 1989, with only ten lots remaining, the society finally called a halt to the program that had operated continuously for seventy-three years.[52]

It did not take long, however, for the phoenix to arise from the ashes. Within ten years of the demise of the vacant lots garden program, a new movement dedicated to transforming "waste places" into productive gardens began to emerge in Edmonton. The community garden movement bore important resemblances to its predecessor; it was essentially an urban movement, it was driven by overlapping, but compatible, agendas, and Edmonton was certainly not the first city in North America to set up an organization dedicated to promoting and supporting community gardens. However, while the vacant lot gardening movement was impelled forward in Edmonton during the First and Second World Wars by the country's appeal to patriotism, the community garden movement took shape gradually in response to emerging social, cultural, and environmental concerns. Edmonton in 1990 was not what it had been in 1920; it was larger, less homogeneous, perhaps more divided along income lines, more environmentally conscious in certain quarters, and certainly less skilled in the gardening arts. Community gardens offered settings in which gardeners could come together to share their interests, socialize, and learn new skills. Whether such gardens were defined by geography, by religious denomination, or by some other criterion, their members often required help to find an appropriate piece of land and then to craft an agreement with the owner so that it could be used for gardening purposes. New community gardens needed start-up assistance with a range of issues—obtaining

Strathcona Rail Community Garden, Edmonton, 2012. [Photo courtesy of Sustainable Food Edmonton]

adequate water supplies, preparing the soil, obtaining and securely storing tools, setting up rules for their gardeners, and designing educational programs for members. Although several community gardens existed in Edmonton in the 1990s, it was assistance rendered to a Mill Woods housing co-op by John Helder, the city's principal of horticulture, that precipitated the organizing of a Community Garden Network to provide support to new gardens and a structure within which all community gardens in the city could flourish.

In 1998, while helping the Mill Woods group obtain permission to set up a community garden on pipeline land, that is land which falls under the jurisdiction of the Provincial Utility Board but is maintained and serviced by the City of Edmonton, Helder discovered the perfect vehicle for facilitating such an agreement. He simply adapted a previously developed vehicle known as the Partners-in-Parks agreement, whereby an individual or a community group agrees to maintain, that is, to plant a garden on, a piece of city-owned property. Suddenly, the Mill Woods project became a model for making publicly owned land available to groups for community (or allotment) garden projects.[53]

The Edmonton and Area Community Garden Network did not spring into being fully formed, said Susan Penstone, who worked as the network's first facilitator and author of its resource manual. Rather, it evolved over a period of years, coalescing around the efforts and beliefs of key individuals such as Catherine Duchesne, who chaired the network's first board, and John Helder. Penstone pegged 1998 as the founding date, because it was in that year that eleven community gardens came together to define the work of a network. From that tentative beginning, the network moved to full incorporation as a nonprofit organization in 2003.[54]

While the vacant lot garden movement appealed to Canadians' patriotism, their belief that by growing a vegetable garden they could participate in a shared, countrywide effort to produce and conserve food stocks, the contemporary community garden movement appeals to their desire for social change. Community gardening is recognized by many as a way to create connectedness and a strong sense of community as an alternative to acceding to the disconnectedness, isolation, and sense of powerlessness that so frequently accompany urban living. Environmental awareness; reconnecting with nature and with the soil; protecting green spaces, thereby improving the atmosphere; contributing to city beautification; protecting local supplies of fresh food; socializing with neighbours; and putting homegrown produce on the dinner table play varying roles in motivating citizens to become community gardeners. Penstone, who cofounded an organization known as Voices of the Soil to promote the concept that the health of the soil can be taken as a direct indicator of the health of a community, was perfectly suited to representing the values of the fledgling Community Garden Network. For her, and for the network, gardening was regarded as an agent of personal transformation, bringing its practitioners into a healthy balance with their environment.

The Edmonton and Area Community Garden Network lasted less than a decade as an independent entity. In 2009, in a move that recalls the much earlier amalgamation of the Vacant Lots Garden Club with the Edmonton Horticultural Society, the Community Garden Network amalgamated with the organization that became, in 2011, Sustainable Food Edmonton.[55] An important distinction between the two amalgamations is that while the horticultural society's mission was built around city beautification, Sustainable Food Edmonton's purpose is to offer community-based programs that focus on local food security, food policy, and food literacy.

Since the First World War, urban agriculture has been a tool in the hands of Canadian governments to deal with food crises of one sort or another; historically, it has been put to use during times of war and economic depression. National policies developed to promote urban agriculture have generally been considered successful, as measured in quantities of vegetables produced, but they have succeeded by drawing on the goodwill and energy of municipalities, individuals, and nonprofit groups. Municipal governments and organizations like the Edmonton Horticultural Society have been eager to respond, partly because they have endorsed the political agenda advanced by the national government and partly because urban gardening of any kind has furthered their own agendas, agendas such as city beautification, weed control, and promoting horticulture as an act of citizenship. However, in the absence of a nationally articulated twenty-first-century food crisis, the job has been left to cities to decide how urban agriculture fits in with their plans for growth and what crisis, if any, they will respond to. Is the security of Edmonton's food supply challenged by environmental or other threats? What are the dangers of creating an urban environment hostile to the practice of horticulture or unfit for urban agriculture? Are vacant lots the only waste places in cities, or are public parks, rooftops, balconies, and walls also suitable for horticultural or agricultural uses?

The process of urbanization is complex. Agricultural land uses give away to buildings, roadways, and recreational facilities. Horticulture takes over from agriculture and food production drops back as a primary use for the spaces we call gardens. But one thing is certain: when waste places are needed to put food on the table, we put aside the notion that gardening is an individual and personal pursuit. Suddenly, as did Mrs. Plugger and her reluctant but ultimately successful spouse, we understand that gardening is a political act.

7 Edmonton, the Rose City

He began to experiment with roses because the conventional opinion was that they couldn't be grown on the prairies. Sure enough, following instructions from books, the first ten he planted died and so did the ten he planted the following year...[O]ne Sunday morning in the 1950s...he turned on the television and watched a gardening program that originated in Winnipeg. The topic was roses. Picking up ideas from that program George tried again and four of the ten varieties he planted survived...[He] went to the Morden Experimental Farm and began to plant their shrub roses which all survived. He was hooked.

—GEORGE SHEWCHUK, Interview

GLADIOLI had been George Shewchuk's major floral interest and he had grown fields of them on his property in Lamont, the small town northeast of Edmonton to which he moved in 1957 when he took up a position there as district agriculturist. He decided to take on a new challenge. First-time failures with roses, combined with tentative successes and a dearth of how-to material on cultivation techniques, set Shewchuk on the path that led, eventually, to his becoming the acknowledged rose expert in the Edmonton area. In 1966, he moved from Lamont to Edmonton, commuting back and forth to work each day. In 1976, he retired to work in his garden and to write a succession of books on rose growing. Books on roses published by Shewchuk in 1981 and 1988 were followed in 1999 by *Roses: A Gardener's Guide for the Plains and Prairies*, which Harry McGee, Ontario rose enthusiast and editor, was to call his magnum opus. In it he set out to show how roses, both hardy and tender, could be made to thrive in the Edmonton area. And Shewchuk practised what he preached. "My Edmonton garden," he wrote in his "1998 Review of New Roses" for the *Rosebank Letter*, "represents what is possible for all Prairie gardeners."[1]

Shewchuk fancied himself a pioneer in the field of rose growing, a groundbreaker. Only dimly perceived by him, because it had disappeared from public knowledge as fast as a flourishing garden declines into a condition of weedy neglect, or vanishes altogether under concrete, Edmonton already had a rose history. It was a history that began long before European settlement, when wild roses grew prolifically along riverbanks and in clearings. But it was also a history that reflected the fascination that so many of Edmonton's early settlers had with the cultivated rose varieties they had come to know in their places of birth—towns and cities in England, France, and even Ontario. As with so much of Edmonton's gardening history, it is a history involving a dismissal of the native in favour of the non-native. *Rosa acicularis*, the prickly wild rose with delicate pink spring blooms so favoured by bees, whose branches were said to have been strewn by Aboriginal people around the home of a recently deceased person to keep that person's ghost from returning, was certainly not enough to make Edmonton a rose city, despite its being chosen in 1930 as Alberta's floral emblem.[2]

Rosa acicularis, one of three wild rose species commonly found in the Edmonton area, was very likely the "wild rose" referred to in articles published in the *Edmonton Bulletin* in the 1880s and 1890s, articles written primarily to attract settlers.[3] A deciduous shrub, it can grow up to five feet tall and its stems and branches, both of which are covered with

Rosa acicularis *or prickly rose. [Photo courtesy of the author]*

prickles and thorns, have given rise to its common name, prickly rose. Fragrant pink single blooms, whose petals have many practical uses in both their fresh and dried forms, come out in May and June. Flowers wither to form scarlet hips that then become food for deer, moose, rabbits, and coyotes, if they are not picked by humans to be turned into teas and jellies.[4] This was the rose that charmed settlers even as they came to value it less highly than the roses they planted hopefully in their gardens, roses brought or ordered from other places.

Despite the reluctance of Edmonton's first gardeners to consider the wild rose for their gardens, its presence in nature was widely perceived to be advantageous to the region. "[T]he most common flower is the wild rose," claimed an article in the July 8, 1882, edition of the *Bulletin*, an article designed to entice settlers with its only slightly extravagant claims for the quality of the soil and the abundance of local fruits and flowers. The emergence of rose blooms in late spring was one of many cheerful markers from nature relied upon by the *Bulletin* to chart the progress of the seasons.[5] And, in 1888, the excellence of Thomas Henderson's honey was attributed in part to the wild rose:

> *Strange to say although [a] quantity of buckwheat has been grown to supply them with honey [the bees] do not go near it finding abundance in the wild flowers and preferring them to the buckwheat. The honey has a slight taste of the wild roses which grow so profusely here.*[6]

It was no surprise, then, that in 1930, when Alberta introduced its Floral Emblem Act, the wild rose won out over the orange lily, a choice explained by P.D. McCalla, horticultural specialist with the Alberta Department of Agriculture, in an article for the *Alberta Horticulturist*. "The wild rose," he wrote, "makes a strong appeal to people interested in native plants"; "[t]he simplicity of the flower, pleasingness and delicacy of form, color and texture and sweet fragrance, are attributes which symbolize the charm, grace and beauty of nature."[7]

Pleasing as prickly rose may have been to many Albertans, its value in 1930 may have depended less on its charms than on its hardiness, the trait that made it, and other native species, useful in the breeding programs that were already being carried out both by amateur gardeners and by horticulturists working in government research stations. But that is jumping ahead of the story. We are still back with Henderson and other early settlers who dreamed of the domestic roses grown in the gardens of their youth.

Evidence of an early interest in growing cultivated varieties of roses begins to show up in the *Bulletin* in the 1880s. In 1885, for instance, the Renfrew Fruit and Floral Company of Arnprior, Ontario, began to advertise that "roses by mail" were a "specialty" of the company.[8] A few years later, it was reported that Thomas Anderson, Crown timber agent, had "got up some cuttings of rose bushes...from his former home in Point Levi Quebec" and that they had "grown remarkably well."[9] But it was Peggy, the wife of Thomas Henderson, who, if her frequent mentions in the *Bulletin* are anything to go by, may have been the first serious rose enthusiast in Edmonton. In July 1890, "Mrs. Thos. Henderson of Fraser Avenue" was reported as having produced "a full blown yellow rose, grown in the open garden." Early the following month, she was reported as having "garden roses" in full bloom and, at the end of August, it was her climbing rose that received special notice. Nor did she lose her touch when the family moved from Fraser Avenue (98th Street) to a farm at Rabbit Hill. In July 1893, the paper reported that "Thos. Henderson of Rabbit Hill brought the *Bulletin* a rose in full bloom last week, which was grown in the open garden." In July 1895, Mrs. Henderson was reported as having "a yellow Persian rose in bloom." The article went on to say that it was an outdoor shrub that had flourished in both Edmonton and Rabbit Hill for seven consecutive years, "so it may be considered to be perfectly hardy in this climate." The fact that two weeks later Mrs. Henderson was also able to produce unnamed pink roses grown on a bush suggests that her interest in roses was more than casual.[10] Where Mrs. Henderson obtained these roses—the yellow of which was likely *Rosa foetida persiana*, a hardy species rose of unknown origins—appears not to have been newsworthy; for the first few years of the twentieth century, the *Bulletin*'s interest shifted first to Donald Ross, and then to Walter Ramsay, when these two men began producing cut roses for sale from their greenhouses. As early as January 3, 1903, for instance, the *Bulletin* reported that Ross had carnations, roses, and fuschias in bloom for cutting in his greenhouse.[11]

The formation of the Edmonton Horticultural Society in the fall of 1909 played an important role in bringing together individuals with similar horticultural interests, interests that could be more successfully advanced co-operatively than individually. Some of these, such as the shared interest in city beautification, pertained to improving the quality of life in Edmonton, but others grew out of horticultural preferences for a particular flower. Sweet peas were great favourites of many early settlers, for example, and gladioli were championed by others. But roses, which were always much more difficult to grow in Edmonton than either sweet peas or gladioli, had more than their share of enthusiasts.

Horticultural shows sponsored by the society reflect the waxing and waning fortunes of rose growers, beginning in 1910. The society's annual show in August of that year included a display of roses that were said to have been grown outdoors by Avonmore Nurseries, and subscription in the rose classes was, apparently, surprisingly high. It "must have been a record year for this King of Flowers," wrote the society's secretary in a promotional issue of *Town Topics* published in 1913, "because at no show since has there been anything approaching the 50 blooms then exhibited."[12] In August 1911, however, the Strathcona Horticultural Society held its annual show and roses, of which there were a "goodly number," were again reported to have "attracted a good deal of attention." Among the "standard varieties" referred to in the article, only one rose was named, the "ever-present American Beauty," a puzzling reference as this rose, which was bred in France in 1875 and made popular in the United States a few years later under the name 'American Beauty', is a tender hybrid and not hardy to the Edmonton region.[13] However, with the arrival in Edmonton of rose enthusiasts Alfred Pike, in 1910, and Ernest Stowe, in 1913, both of whom became involved in the activities and direction of the horticultural society, interest in rose growing increased and the technique of grafting tender varieties onto hardy rootstock expanded the repertoire of aspiring growers. It was with the help of these two men that, two decades later, rose enthusiast Walter Wilson set out to "beat the odds" and make Edmonton the "Rose City."[14]

Signs that Edmonton would respond favourably to Wilson's rose campaign began to appear in newspapers in the 1920s and early 1930s, culminating in 1932, the year before Wilson and Pike launched their Capitol Theatre show. In May 1924, for example, the *Bulletin* published an article on vines and shrubs recommended by the experimental station in Beaverlodge for planting in the Edmonton area; *Rosa rugosa*, a hardy Japanese species rose being used in hybridization experiments, was among the station's recommendations.[15] Three years later, in 1927,

Mrs. Blyth in her garden, Edmonton, September 1937, with the John Lefeuvre rose trophy she had won that year in the Edmonton Horticultural Society annual horticultural show. John Lefeuvre, a former society board member, donated this trophy to the horticultural society, and it was first awarded in 1928. [PAA BL 215/2]

roses were singled out for attention by the *Edmonton Journal* in its report on that year's fall horticultural society show. Not only had roses made the "most striking advance of any variety," according to the reporter, but the one hundred blooms entered in the 1927 show were far and above numbers entered in previous shows.[16] The following year, in 1928, the John Lefeuvre challenge trophy for roses was awarded for the first time at the August show. It was won by a Mr. H.A. Holland, who was featured in the *Bulletin*'s account of the show as a big winner and whose prizes included that for the "best individual rose in the show"; Holland's expertise in rose growing was transferred to readers of the *Bulletin* the following spring in an article headlined, "Start Right If You Want Good Roses."[17] Nurseries were quick to sense a market; the 1932 prize list published by the Edmonton Horticultural Society included an advertisement by "Wm. Ferguson," of Dunfermline, Scotland, for Ferguson's Scotch Roses. Not only did Ferguson's claim to sell the "cheapest" roses and those that would best survive the winters; it also claimed that roses sold by the company in previous years had been winners in the annual horticultural shows and that its 1932 orders were already in excess of one thousand bushes. Ferguson's had four hundred varieties to choose from and it had an Edmonton-based agent.

Edmonton gardeners with a keen or a passing interest in rose growing were given ongoing support and information by H.W. Stiles, a rose enthusiast who had served as the society's president in 1930 and whose weekly columns on gardening for the *Bulletin* in 1932 paid particular attention to roses. In addition to offering an abbreviated history of his favourite flower, Stiles recommended planting hybrid perpetuals for their relative hardiness in Edmonton and hybrid tea roses for their repetitive blooming. He reported on the two dozen bushes from Scotland he had planted in his own garden, presumably, but not explicitly, acquired from Ferguson's of Dunfermline. To keep pests at bay, he advocated his own practice, which was to spray three times a week with a "nicotine solution." And Stiles did not confine his columns to his own

> *Walter Wilson in his garden, Edmonton, EB, June 8, 1949. [CEA EA-600-2510]*

experience as a rose grower. His August 9 column that year drew attention to the roses of a Miss Isobel Holme, whose garden featured a hedge with 125 roses in it, "among which a little white plaster bunny snuggles cosily."[18]

Wilson, referred to as Edmonton's "Mr. Theatre" when he retired from his position as manager of the Paramount Theatre in 1954, was not the most likely of persons to conceive and direct a campaign to make Edmonton known for its roses. Unlike his associate in the scheme, seedsman Alfred Pike, who sourced and sold vast numbers of rose bushes over a period of several years, and provincial gardener Ernest Stowe, who judged almost all the annual rose shows, Wilson did not work in the field of horticulture. Rather, he spent his entire working life in the city as the manager first of the Capitol, and then of the Paramount, movie theatres. Born in Yorkshire, England, in 1876, Wilson was close to thirty-five years of age when, in 1909, he moved to Winnipeg and began a career in what must then have been the new world of moving pictures. In 1923, he settled into a house on 118th Street just south of Jasper Avenue, a house now replaced by multi-storey residential towers built to take advantage of the spectacular views south, looking over Victoria Golf Course and the North Saskatchewan River. It must have been there that he began to indulge his passion for growing roses, a passion he was eager to share with others. In 1928, he began broadcasting three radio programs a week on gardening for the local radio station CJCA, doling out advice to gardeners, much of it on roses, in his heavy Yorkshire accent and billing himself as the "world's worst radio announcer."[19] His granddaughter Barbara Paterson's most vivid memories of her grandfather are of the broadcasting equipment set up in his office at the Capitol and of his heavily accented voice coming over the radio. Every August, she remembered, her grandfather would put together a bouquet of roses for her father to deliver to his wife on their wedding anniversary; a rose for each year her parents had been married. In 1933, by which time he was approaching sixty years of age, Wilson put his management skills to work to shift the city's horticultural horizons. "Make Edmonton the

Rose City," read an advertisement in the *Bulletin* on July 17, 1933, and, for the next twenty years, through will, persistence, and collaboration, he strove to achieve his goal.[20]

The first Capitol Theatre Rose Show took place on July 24, 25, and 26, 1933, in the lobby of the Capitol Theatre, and was attended by as much fanfare as Wilson and the *Bulletin* could create for it. Judge Stowe claimed it to have been the "best display of roses ever seen in this city." The theatre lobby was crowded with admiring spectators on opening night. And the show's promoters, Wilson and Pike, cultivated an even wider enthusiasm for growing roses by giving away 125 rose bushes to audience members, "those who had birthdays or were otherwise celebrating." To show their appreciation for those who entered the first show, Wilson and Pike placed a thank-you advertisement in the *Bulletin*.[21]

If the purpose of the Capitol Theatre Rose Show was partly to prove that roses could be successfully grown in Edmonton and partly to stimulate an interest in rose growing, the first decade of its existence was promising. Six thousand "Edmonton growers" were said to have taken part in the second annual show, which took place on July 23 and 24, 1934, a number that was doubtless based on the reporter's enthusiastic, if faulty, assumption that the number of bushes sold by Pike translated into an equal number of competitors. The rose judged to have been the best in show that year, a crimson red hybrid tea by the name of 'George Dickson', would have required special care to thrive in Edmonton's climate, but it proved to be a recurring winner, taking the prize for best rose in show in 1937, 1941, and 1944 as well. "To those who sigh over the gardens of west coast cities," the report in the *Bulletin* read in 1934, "the display of roses grown in Edmonton is indeed promise of a garden city on the prairies." Spurred on by this success, Pike was said to have had ten thousand rose bushes ready to sell to exhibitors and to aspiring growers in 1935.[22]

The 1941 catalogue published by Pike offered an impressive choice to rose growers, helping to explain why Edmonton may have reached its apogee as a rose-growing city during the first half of the 1940s. Pitched to local conditions, the catalogue highlighted the fact that all bushes included in it were "grown on a Briar the most suitable to withstand our severe winters and variable climatic changes" and that all varieties had been "intensively tried out and proved as to their suitability for Edmonton and district." Forty hybrid teas were listed at fifty to seventy-five cents each. There were four hardy perpetuals, three climbing roses, including the popular buff-coloured 'Gloire de Dijon', and eight double rugosas, including the very popular 'Hansa' rose and the Persian yellow species that Peggy Henderson may have grown back in the 1890s.[23]

However, if seedsmen and nursery owners responded with entrepreneurial vigour to Wilson's "Rose City" slogan, and if the Edmonton Horticultural Society contributed expertise, advertisements in its annual prize lists, and a receptive membership of experienced gardeners, responsibility for the overwhelming success of the venture must, in the end, go to the theatre manager himself and to his inveterate media sponsor, the *Edmonton Bulletin*. Newspaper coverage of the shows generally began the Saturday before the Monday opening, coverage that included a photograph of the challenge cup donated by the *Bulletin* for the grower of the bloom judged best in show. The accompanying text would invariably include the assertion that the show had done much to encourage the growing of roses generally and that it had validated the mission of the founders by making Edmonton a "Rose City." In 1944, this Saturday coverage took the form of a full-page advertisement, artfully composed to catch the attention of readers without sacrificing the message that Edmonton had achieved its founder's civic goal by becoming a city recognized for its roses.[24]

The impressive and well-patronized series of shows staged at the Capitol in the 1940s appear to have gone some way towards fulfilling the dreams of their founder. Despite an excessively dry summer in 1941, which meant a slight diminishment in the quality of the entries, the quantity remained high, numbering something between two and three hundred exhibits. Wilson used his time behind the microphone to look forward to 1942, when the show would be celebrating its tenth anniversary. He expected this tenth show to "mark a milestone in the endeavors of so many enthusiasts to make Edmonton the Rose City" and he was not disappointed.²⁵ The year 1942 brought perfect rose-growing weather; more than four hundred exhibits were crammed into the repurposed theatre and J.C. Bowen, Lieutenant-Governor, "was the guest speaker on a short broadcast from the lobby." Judge Stowe was delighted by the overall high quality of the exhibits and had only good things to say about the 'Earl Haig' rose that won the challenge cup for its grower John Home. Further, the *Bulletin* reporter believed that Edmontonians would be very interested to hear "the words of congratulation offered by American visitors who were agreeably surprised at the quality of the roses for a city this far north."²⁶ Ideal rose-growing weather persisted through 1943, bringing yet another improvement in the "variety and quality of lovely blooms." Prize winners all took home a little something in addition to their ribbons, a bottle of apple blossom cologne from the Corner Drug Store, perhaps, or a "Gentleman's Hat from Modern Tailors." The *Bulletin* published the list of merchants who had contributed items from their inventories to be given out as prizes.²⁷

By 1950, the Capitol Theatre Rose Show was firmly linked to the indomitable character of its founder and patriarch. On July 15 that year, an advertisement for the show claimed that "[t]he Capitol rose growing campaign has been the largest city-wide beautification effort ever staged in Canada." Wilson was interviewed for an article that appeared two days later. He had been "dismayed" upon his arrival in Edmonton in 1923, he was quoted as saying, "to find great horticultural art restricted to only a few Old Country people." The "Rose City" campaign had been launched only after Wilson had satisfied himself that "great horticultural art," a concept inextricably linked in his mind to the cultivation of roses, was achievable. The technique of burying roses in the fall, watering them in heavily, digging them up again in the spring, and replanting them, a technique referred to by Wilson as the "simple solution" to the problem of cold winters, had become accepted practice for rose growers. All in all, the reporter suggested, the success of the campaign to make Edmonton the "Rose City" was primarily attributable to Wilson's vision and to his efforts.²⁸

Hilda McAfee, Edmonton, EB, July 18, 1949. [CEA EA-600-1082e]

In Hilda McAfee, who won five of the six open competition awards in the 1950 rose show and who dominated the prize winners during the last few years of the show's existence, Wilson met a woman whose combination of industry in the garden and high social tone must have matched his conception of the ideal competitor. McAfee did not take home the *Bulletin* cup for the best rose in the show in 1950, but she made up for it the following year by winning it with her 'Peace' rose, described in the *Bulletin* report as a "product of the horticultural artistry of Mrs. J.A. McAfee."[29]

Unlike the majority of prize winners in the Capitol Theatre Rose Show, most of whom emerged briefly from the anonymity of their homes and gardens to accept their accolades before dropping back into obscurity, McAfee thrived on the celebrity she earned as a gardener. "Hilda's specialty was roses," her nephew Ernest Hodgson wrote in biographical notes he provided to the City of Edmonton Archives, a specialty that resulted in her being referred to as the "Rose Queen of Alberta." An inveterate competitor in horticultural shows throughout the 1940s and 1950s, McAfee won so many prizes that eventually, according to her nephew, she ceased entering competitions and instead offered cups and prizes for others to win. In August of 1950, as a particularly successful horticultural season drew to a close, McAfee declared jokingly to a newspaper reporter that her husband steadfastly refused to polish the cups she had won so she was forced to pay for the pleasure of winning by doing the polishing herself.[30] An article about her that appeared in the May 1957 issue of *Canadian Homes and Gardens* claimed that McAfee had won five trophies and eighty-five prizes in shows in 1954 and an equivalent number of cups and prizes the following year, a record that perhaps partly accounts for her having been named, in 1955, a "fellow" of the Royal Horticultural Society, Britain's largest and most influential gardening organization.[31]

Born in Stoke-on-Trent, England, in 1900, Hilda May Cheetham came to Edmonton in 1912 with her family. Her gardening career began in August 1930, when she married Jim McAfee, moved to a small bungalow at 11330–92nd Street, and set about to transform the 33 by 120 foot lot

Hilda McAfee accepting an award from Walter Wilson for her 'Peace' rose at the Capitol Theatre Rose Show, Edmonton, July 1951. [CEA EA-507-42]

into a paradise of plants. "It was amazing," wrote Hodgson, "that so much beauty could be organized to fit that small thirty-three foot city lot."[32]

McAfee gravitated to the media spotlight whenever an opportunity presented itself. Photographs of her and of the garden she built with the help of her husband Jim appeared frequently in Edmonton's daily newspapers and in magazines such as *Canadian Homes and Gardens* and *Canadian Horticulture and Home*.[33] In all these photographs McAfee is dressed as if for a special occasion and she is beaming radiantly into the camera. In 1951, after winning the prize for having grown the best rose in the Capitol Theatre Rose Show, she was photographed for the *Journal* as she was being presented with her trophy. Head thrown back and tilted slightly towards the camera to show off her smile and holding the "delicately fashioned yellow flower with pink tints of the 'Peace' variety" that had taken the prize, she was shaking a somewhat stiff, but smiling, Wilson by the hand. It is an image that marks the end of an era in Edmonton's rose history.[34]

Wilson and Pike may not have succeeded in their effort to brand Edmonton forever as the "Rose City," but, at the time of Wilson's retirement in 1954, they could be forgiven for having imagined it possible. Pike & Co.'s 1954 catalogue listed eight hardy rugosa varieties, five hybrid perpetuals, and eighteen hybrid teas, a selection of hybrid teas that was to decrease in the years immediately after the last year of the campaign. And Pike was not the only local seed and nursery business to offer a rich selection to rose growers. The H. Arends' Rose Gardens catalogue that year listed twelve hardy shrub roses, including 'Betty Bland', 'Hansa', and 'Karl Forster', and twenty hybrid teas. The latter were listed by colour and included the popular 'Peace' rose that had won Hilda McAfee a prize in 1951, a variety described as having unusually large and long-lasting blooms of "lemon yellow" colour with outer petals "edged cerise or apple-blossom pink."[35] Horticultural society support for the rose campaign appeared to be solid. A meeting of the society scheduled for April 22, 1954, for example, promised "a very special showing of the H.M. Eddie Rose Gardens in Vancouver," the company from which Pike & Co. purchased its rose stock. The meeting was to include a presentation from Wilson, who "will give us the benefit of his years of experience of rose growing in Edmonton," and Hilda McAfee was conscripted to "bring greetings to the ladies."[36] It must have seemed, in 1954, that the rose city campaign was unstoppable.

Stowe retired from his job as chief gardener for the Province of Alberta in 1952 and moved to British Columbia, a few years before the rose shows ceased being held.[37] McAfee, who won the prize for the best rose in the last of the annual rose shows in 1955, died just before her sixtieth birthday in June 1960, leaving her husband, Jim, to tend the garden upon which she had lavished so much care and attention. Wilson died nine years later, on March 27, 1969, at the age of ninety-two, leaving behind an

already fading memory of the campaign to make Edmonton a city known for its roses.[38] And although Pike continued to run his business until his death in November 1981, aged ninety-six, he appears not to have capitalized on the new wave of rose culture that began its slow roll across the Prairies as early as 1950, the year 'Thérèse Bugnet', a locally bred rose, was registered. Although they could not have known it at the time, Edmontonians were moving into a new era of rose growing, one dominated not by celebrity exhibitors at horticultural shows, whose passion for rose growing consumed much of their leisure time, but by hybridizers, whose successes in breeding roses that were adapted to the conditions of the northern plains and Prairies made rose growing a feasible proposition, even for the casual gardener.

The 'Thérèse Bugnet' rose revolutionized rose growing not only in Edmonton but across the Prairies.[39] The first locally hybridized rose in what became an ongoing stream of introductions by both amateur hybridizers and government research stations, 'Thérèse Bugnet' offered gardeners a chance to grow showy double pink and fragrant flowers that bloomed continuously from May to September on a four-to-six-foot shrub, the dark red branches of which showed up well against the winter snow. Best of all, perhaps, it was a rose that did not require the elaborate care given to tender roses by so many of Edmonton's rose enthusiasts. It would be fanciful to suggest that 'Thérèse Bugnet' owed any part of its popularity to its association with its hybridizer, whose horticultural triumphs seemed so inextricably linked with his powerful personality, but it is difficult to avoid the speculation. Few of the hybrids developed by amateur hybridizers have maintained their place in the market as has this particular rose, and in Edmonton gardens at least, its inclusion is sometimes a deliberate reference to the local.

Claude Roberto is one Edmontonian who has made a point of including a 'Thérèse Bugnet' in her Edmonton gardens. For Roberto, who was born in Marseille, France, and who came to Edmonton in 1977 to attend the University of Alberta, having this particular rose in her

< *Ernest Stowe, provincial gardener, Edmonton, 1952. [PAA, PA 950/1]*

garden has come to have an almost mystical significance, symbolizing the link between her birthplace and her adopted home. For Roberto, gardening is a meditative and restorative activity. Much as Bugnet was motivated by the desire to bring to Canada a beauty equivalent to what he had known in France, so Roberto takes her inspiration in gardening from the idea that the beauty she creates will bring others as much enjoyment as it brings her. Gardening, for Roberto, brings about a sense of connectedness with nature and with the entire universe, a mystical way of thinking that is reminiscent of Bugnet's own expressed awe before the "magnificently creative power which some call God, and some others Nature, meaning, after all, the same thing, the same unfathomable entity."[40]

Rosa acicularis, or prickly rose, commonly but mistakenly thought to have played a key role in the genetics of 'Thérèse Bugnet', was only one of several native rose species to be incorporated into the breeding programs of both amateur plant hybridizers and their counterparts working in government research stations and universities. *Rosa arkansana*, also called the low prairie rose because of its small size in comparison with other rose species native to northwestern Canada, was the "parent of all Parkland roses," according to H.H. Marshall, who began to develop them in 1957 while he was "Head Gardener" at the federal experimental farm in Brandon, Manitoba. This breeding program, carried on by Marshall and his successors from 1957 to 2010 at the Canadian government research stations in Brandon and Morden, Manitoba, resulted in over a dozen introductions of hardy roses. *Rosa woodsii*, or Wood's rose, widely distributed around western North America, was integrated into the breeding programs of many hybridizers, and *Rosa blanda*, native to Canada but not to the Prairies specifically, was valued by hybridists as the only native species totally lacking in thorns and prickles. These two are now thought more likely than *Rosa acicularis* to have played a role in the parentage of 'Thérèse Bugnet'.[41]

The mystery that developed around the parentage of 'Thérèse Bugnet', a mystery centred on the role that *Rosa acicularis* did or did not play in the breeding, is of interest less because of its particulars, which take us deep into the intricacies of plant genetics, than because it points to the difficulties encountered by amateur hybridizers whose work, each step of which requires meticulous documentation, extends over many years. Approximately twenty-five years elapsed between the beginning of Bugnet's rose hybridizing experiments and the official "coming out" of 'Thérèse Bugnet'. And, although it remains the most well known of his roses, having captured both the imagination of gardeners and a sizable share of the market for hardy shrub roses, other Bugnet rose hybrids, including 'Louise Bugnet', 'Madeleine Bugnet', 'Marie Bugnet', 'Lac La Nonne', and 'Lac Majeau', can still be found in public gardens and in nurseries. As a group, they represent a lifetime of work, each step of which was recorded by Bugnet in notebooks in the cryptic and sometimes abbreviated style that required decoding before the parentage puzzle of 'Thérèse Bugnet' could be solved. As a group, they also represent a significant horticultural achievement, one that brought him recognition before his death and continues to do so.

While Bugnet has received ample recognition for his contributions to what the great American author, editor, and rose specialist Horace McFarland referred to at least once as "the march of the rose," the Edmonton plant hybridist Robert Simonet has not. Remembered instead for his overall contributions to the development of horticulture in the Canadian northwest, and more particularly for his work with double flowering petunias, Simonet's influence on the development of hardy roses has been assessed by Paul Olsen as far outweighing his reputation in this area. Unlike his fellow amateur hybridists, including Bugnet in Rich Valley, Wright in Saskatchewan, and Frank Skinner in Manitoba, Simonet's primary interest was not hardy shrub roses. The two rugosa shrub roses he is best known for, 'Simonet's Double Pink' and 'Simonet's Thornless', failed to gain a market presence, although the latter, writes Olsen, is "the only thornless rugosa hybrid ever developed." Instead, according to Olsen, Simonet "made a bold attempt" to breed cold-hardy hybrid teas. Although, after approximately twenty-five years spent in the attempt, Simonet's hybrids failed to achieve that perfect combination of cold hardiness with the floral qualities of hybrid teas, two of his selections became integral to the development of the Agriculture Canada Explorer series after they were obtained by Dr. Felicitas Svedja during a 1964 visit to Prairie rose breeders. "[F]rom [Simonet's] innovative program breeding Hybrid Teas with species," Olsen writes, "the Agriculture Canada Explorer cultivars of *Rosa kordesii* were eventually developed." Thus, despite Simonet's failure to achieve the goal he set for himself, his work managed to find its way into the gardens of Edmontonians by an indirect route.[42]

The 1960s, and the decades that followed, saw Edmonton gardens transformed by the steady stream of hardy rose cultivars developed at federal agricultural research stations in Manitoba, under the leadership of Henry Heard Marshall and his successors (Parkland Roses), and in Ontario and Quebec, under the leadership of Dr. Felicitas Svedja and her successors (Explorer Roses). 'Assiniboine', the first in the Parkland series, came out in 1962, while the first of the numerous Explorer roses, a soft pink rose named after explorer Martin Frobisher, came out in 1970. By 2002, forty new hardy roses had become available to gardeners in these two series alone to add to cultivars developed by amateur hybridists such as Bugnet, Wright, Skinner, and others. While gardeners with a preference for hardy roses had formerly been limited to shrub roses originating in Europe and Asia, they now had choices that had been bred for Canadian climates, some using Canadian species, and that bore names associated with Canadian history and geography. The interest aroused within Edmonton's horticultural community by the introduction of so many hardy roses sparked a new phase in Edmonton's history as a rose city, a phase marked simultaneously by the increased use of roses as hardy landscape choices, both for public parks and for private gardens, and by a revival of interest in tender roses by passionate amateur gardeners. If rose growing in Edmonton had become much easier by the 1970s, it had also become more complicated; the *Alberta Horticultural Guide* published by the provincial Department of Agriculture in 1978 listed eight species roses, twelve tender roses, and twenty-four hardy shrub roses, including

'Thérèse Bugnet', 'Marie Bugnet', and releases from the Parkland and Explorer series. It was to the combination of new possibilities and old constraints that Shewchuk addressed himself when, in 1976, he retired from his job with the provincial government and turned his full attention to roses.[43]

Shewchuk's retirement career as a garden writer and educator arose naturally from his previous interests and training; an early interest in plants and gardening had been followed first by a period teaching school and then by a career as a district agriculturalist. By turning to hobby gardening full-time after his 1976 retirement, he was able to bring his interest in roses together with his professional training and experience. Although the means he employed were very different from those employed by Wilson before him, his goals were similar; Shewchuk wanted to inspire and educate gardeners so that more roses would be grown in Edmonton and its surrounding communities. No hybridizer himself, Shewchuk admired and followed the work of local hybridizers such as Simonet, Bugnet, and John Wallace at Beaverlodge, Alberta, and he grew their roses in his own garden. His connections with the horticultural community, many of them made during his years as a district agriculturalist, proved useful. As he continued to add both tender and hardy roses to his own garden, gradually turning it into a showplace containing up to 350 different roses, many of them hybrid teas, he assumed new relations with organizations and institutions such as the Edmonton Horticultural Society, to whose members he spoke frequently on topics related to rose growing; the Faculty of Extension at the University of Alberta, where he taught courses on rose growing; Olds Agricultural College, where he taught courses on judging roses in horticultural shows; Lakeland College in Vermilion, Alberta, where he taught courses on both rose growing and judging; and the University of Alberta Devonian Botanic Garden, where he taught some classes on rose growing and led tours of rose gardens in Edmonton. For approximately twenty years, beginning around 1980, Shewchuk's garden was included on a list of demonstration gardens nominated by the Canadian Rose Society. And, in 1996, Shewchuk was referred to in an issue of the Ontario-published *Rosebank Letter* as "Edmonton's ranking dean of roses," a position confirmed three years later when his final and most comprehensive book on rose growing was published by the University of Alberta's Faculty of Extension. He may not have been the promoter that Wilson had been in the first half of the twentieth century, but, through his writing and teaching, Shewchuk probably did more than anyone else in Edmonton in the latter part of the twentieth century to stimulate and sustain an interest in rose growing.[44]

It was a tour of Edmonton rose gardens led by Shewchuk, perhaps the one advertised in the *Rosebank Letter* for July 14, 1996, that decided Heiko Lotzgeselle to take his casual interest in rose growing to a new level. The tour began at the University of Alberta, Lotzgeselle remembered, and it included the rose garden at the University of Alberta Devonian Botanic Garden, as well as Shewchuk's own home garden. Although his own garden bears no resemblance to any he saw that day, it was the sight of so many different roses in Shewchuk's garden, all of them integrated into a residential landscape, that inspired Lotzgeselle to convert his own garden from one that just happened to contain roses to a real rose garden. It was as though a seed of interest planted years before had suddenly found soil in which to germinate.[45]

Lotzgeselle and his wife, Carol, had come to Edmonton in 1971, the year he accepted a job with the Alberta Housing Corporation. But it was some time in the mid-1980s, while the Lotzgeselles were living in a house on Ramsay Crescent in the Riverbend area of the city, that he caught his first glimmer of the attraction roses would come to have for him. He and Carol had carved a small accent garden out of a portion of the front yard and included one rose in the plantings. The bloom, when it emerged, was a beautiful rich red colour with a pronounced and lovely fragrance; of all the flowers in the bed, it was the one they most admired. Although he has searched for it since, Lotzgeselle has not found it. But in his own mind there is a link between his memory of that first rose and the decision sparked by Shewchuk's guided tour to take up rose gardening himself.

The approximate ten-year interval between the first rose on Ramsay Crescent and Shewchuk's guided tour of rose gardens saw both Lotzgeselles develop an increasing interest in gardening, generally, and in roses, in particular. In 1991, they moved to a house in a new area of the city, one with a large lot to landscape. They planted some spruce trees for privacy and laid out a gravel border around the perimeter of the lot, dotting it with a few junipers. An accent bed for perennials in the front yard included not one but five or six roses; again, the Lotzgeselles were especially drawn to the roses. Carol enrolled in a course on roses put on by the City of Edmonton to learn more about growing them and then Heiko signed up for the fateful tour. More than twenty years after the move to Reeves Way, and with elements of the first plan still in place, although the gravel border has gradually been replaced by perennial beds, a gazebo, and a water feature, the Lotzgeselles' garden contains more than four hundred different roses along with many other perennials, including a collection of peonies. Hardy shrub roses, including some from the Parkland and Explorer series, are used where they serve a purpose in the overall architecture of the garden, either as background plants or in spots where they climb. Heiko likes the hardy hybrids and is especially fond of 'Winnipeg Parks', one of the Parkland series, but he favours the tender roses, especially hybrid teas and grandiflora cultivars that are not hardy to Edmonton's climatic conditions. These are given pride of place in an interlocking series of rose beds that weave their way around the back garden and into the front. All the roses in these beds require winter protection.

Maintaining such a large and specialized garden takes Heiko Lotzgeselle approximately four hours a day, from the middle of May until the roses are "put to bed" in the late fall. Every year, he replaces plants that die or fail to flourish with new roses, roses sourced and acquired by mail from Palatine Roses in the Niagara region of Ontario, perhaps, or from Brad Jalbert, who sells and breeds roses in Langley, British Columbia, where he runs Select Roses. But the Lotzgeselles also frequent local garden centres for interesting roses and are not above

buying something from a big box store to try. Care includes pruning, weeding, fertilizing, watering, ensuring that the plants are healthy and free of pests and disease, covering them for the winter, and uncovering them each spring. Winter protection is a process that has been subject to much experimentation in the Lotzgeselles' garden. Although Heiko first adhered strictly to Shewchuk's method of trenching, burying, and covering the roses with peat moss, he has also tried fibreglass insulation sandwiched between layers of plastic, styrofoam "rose huts" that are produced commercially for the purpose of protecting roses, and, finally, a combination of rose huts and insulated tarpaulins. This last combination has proven to be the most effective, and also the least labour-intensive, of the various methods. And, finally, there are always new projects in the garden such as relocating a bed or creating a special feature. But, says Heiko, when he and Carol take a break on the weekend, make a cup of coffee, and sit on the swing to survey the garden, it is all worth it.

So, if Edmonton is not "the" rose city, is it "a" rose city, and if so, what could this possibly mean? The Lotzgeselles' garden is only one of perhaps many rose gardens planted and lovingly cared for by passionate amateur gardeners, who, like Heiko, are fascinated by the variety, beauty, and heady fragrances of roses at the same time as they are drawn by the place roses occupy in so many cultures and mythologies. And then, as Lotzgeselle says, no other flower rewards the gardener by continuously blooming from May to late September, depending on the weather. Private rose gardens, however, no matter how beautiful they are, are not publicly visible or accessible. Without an opportunity to emerge from the private sphere and into the public one, the kind of opportunity provided by the annual Capitol Theatre Rose Show, the rose brand for Edmonton cannot be made to stick.

If, on the other hand, the use of roses in public landscaping is anything to go by, Edmonton might warrant some consideration as a rose city. In June 2003, the City of Edmonton opened a public rose garden in the southeast corner of Louise McKinney Riverfront Park, close to

< *Top: Rose garden of Heiko and Carol Lotzgeselle, Edmonton, 2011. [Photo courtesy of Carol Lotzgeselle] Bottom: 'Paradise' rose in Lotzgeselle's garden. [Photo courtesy of the author]*

Edmonton's downtown area. Intended to demonstrate the range and variety of roses that can grow in the Edmonton region, it includes cultivars from the Explorer and from the Parkland series, as well as some of the roses developed locally by amateur hybridists. Roses are an essential element in the landscape design for Mackenzie and Mann Park, which links Oliver Square to the Queen Mary Park area of the city at 113th Street, again close to the downtown area. 'Thérèse Bugnet' roses, planted in hedge formation, are combined with blue metal to create a sculpture at the top of Victoria Park Hill. Rose beds maintained by the Edmonton Horticultural Society at the Muttart Conservatory in the river valley flank the sidewalk leading from the parking lot to the conservatory entrance, and roses are included in the design of the Edmonton Horticultural Society's Centennial Garden in Henrietta Muir Edwards Park. A small public rose garden was developed in 2010 as a central feature of the Peace Garden, a combined community garden and park in the Oliver area of the city. And, finally, hardy roses appear in public beds on street corners or in city parks throughout the city, often indifferently or even poorly maintained but there nonetheless.

In 1964, fewer than ten years after the Paramount Theatre closed its doors on the last annual rose show but just as the hardy shrub roses were coming onto the market, the marigold, a flower that has no species native to Alberta, was chosen to be Edmonton's official flower.[46] Its bright gold and burnt orange colours were intended to represent both "sunny Alberta" and the Klondike Gold Rush, this last of which was a contrived and somewhat dubious attempt to connect the marigold with the city's history. The flower was described as being, like Edmontonians, "of strong and sturdy stock," a questionable if saleable proposition. And, because the marigold has many species and varieties, it was said to represent the diversity of Edmonton's population.

Since 1964, the marigold has featured in floral arrangement classes included in the Edmonton Horticultural Society's annual flower shows, shows that no longer exist. It has been used in annual feature beds planted by the City of Edmonton Parks Department and by home gardeners as a colourful and reliable plant for annual flower beds or ornamental pots. But it is the rose that has haunted the imaginations of Edmontonians since the beginnings of the city's history and the rose that has inspired the creativity, artistry, tenacity, and even genius of its enthusiasts and its hybridizers. Edmonton's floral emblem may be the marigold, but its beating heart is the rose.

8 The Invisible Tapestry

Remembering Edmonton's Chinese Gardeners

As I stand on the stone bridge in Edmonton's Chinese Garden, I wonder if I belong in this city. The dozen Chinese zodiac statues which ring the pond offer little help. The dragon and tiger sculptures look as out of place in Louise McKinney Park as I do.

— MARTY CHAN, "Chinese Garden Reflection"

There below [Hong Lee's garden is] a veritable Chinese tapestry worked out in rectangles of harmonious greens, sea green, sage green, olive green, and the delicate apple green of lettuce beds.

— HELEN GORMAN CASHMAN, *EJ*, August 14, 1935

North Saskatchewan River valley, looking down from the bank at 121st Street and 100th Avenue, Edmonton, n.d. [CEA EA-9-165]

ZODIAC STATUES, double-tiered pavilions, and stone bridges across rock ponds were not featured in the gardens of Edmonton's first Chinese settlers, those who arrived in the late nineteenth and early twentieth centuries. The prime concern of these men, and they were all men, was survival. They came to Canada, many still only boys, with one purpose—to make money to send back to their families in China. They came to Edmonton from points west and south, having arrived in the port city of Vancouver before working their way into Alberta, hoping perhaps to settle where anti-Chinese sentiment was less virulent than it was in British Columbia at the same time. Those who chose gardening as an occupation did so to earn a living. Onions, cabbages, and beets were typical features of the gardens they tended, many of them strung out along the North Saskatchewan River valley from Beverly in the east to Government House in the west.

Were these early gardens in the minds of those who decided, a full century later, to raise money for a traditional Chinese garden to be located in Edmonton? What was the value of such a project, either to Edmonton's Chinese community or to the community at large? The Dr. Sun Yat-Sen garden, opened in Vancouver in April 1986 and immediately recognized as a visible symbol of traditional Chinese culture in the middle of a sophisticated western Canadian city, quickly became both a valued civic landmark and a popular tourist attraction. The Kurimoto Japanese Garden, opened as part of the University of Alberta's Devonian Botanic Garden in September 1990, exposed Edmontonians and tourists to a renowned expression of Japanese culture. It could be argued that both these gardens raised the profile of their sponsoring cultural groups and helped to bridge cultural divides between sponsoring and host communities.[1]

In the year 2000, the Edmonton Chinese Garden Society was incorporated under the Alberta Societies Act and achieved registered charity status. Then Mayor of Edmonton Bill Smith had offered the society a river valley site that it decided to accept, a site on a south-facing hillside that sloped towards the North Saskatchewan River from the eastern edge of the downtown. Louise McKinney Park was, in 2000, in the process of being developed and promoted as a model urban park. The Chinese garden project became part of the development plan.[2]

By accepting the city's site offer, the Chinese Garden Society relieved itself of the burden of land acquisition and gained an ally and partner. Councillor Terry Cavanagh, who sat on the society's board of directors as the City of Edmonton representative, became one of the garden's most ardent supporters and promoters. However, the choice of site, a difficult one even for the society's board of directors, disaffected some members of Edmonton's Chinese community, who would have preferred to incorporate the garden into nearby Chinatown and who were opposed to the concept of an open, unsupervised garden.[3]

In the end, the directors of the Edmonton Chinese Garden Society continued on the course they had chosen. As George Ng, who joined the board of directors in 2000 and became its co-chair in 2003, said, the society had to look at what was realistic. No offer of a suitable downtown site was about to materialize; the river valley site would give the general public exposure to the garden in an open, recreational setting, a fact society directors decided to make an advantage.

The historical rationale for building a traditional Chinese Garden in Edmonton's river valley was not emphasized in the fundraising campaign, but it was articulated by Ng in the business plan he prepared for the society in May 2003. Until the 1960s, he wrote, "the site of Louise McKinney Riverfront Park was utilized by the Chinese to grow Chinese vegetables for local consumption." The planned traditional garden, intended to honour Chinese pioneers who made contributions to Edmonton's development, "will help enhance the understanding and appreciation of the Chinese culture, and have a positive impact on the quality of life of the community."[4] For George Ng, and for others who assisted in the realization of the project, including architect Francis Ng, who was part of a team of people who designed the garden, the Chinese Garden in Louise McKinney Park was built to be a visible symbol of Chinese participation in the ongoing project of city building that began more than a century ago and continues today.

In an article published in the summer 2009 issue of *Legacy* magazine, author Marty Chan, contemplating his own identity issues and their relation to his work as a writer, was first struck by how culturally foreign the Chinese garden appeared to him. "The dragon and tiger sculptures," he wrote, "look as out of place in Louise McKinney Park as I do." Chan was too young to have witnessed for himself the "veritable Chinese tapestry worked out in rectangles of harmonious greens" that so impressed *Edmonton Journal* writer Helen Gorman Cashman, when, in 1935, she overlooked Hong Lee's garden on what is today a part of Victoria Public Golf Course.[5]

Chan would not be alone in his inability to call up images such as the one evoked by Cashman. Hong Lee's garden, and the many other Chinese market gardens that became community landmarks throughout Edmonton, especially in the first half of the twentieth century, have no place in the recorded history of the city. All that remains as evidence is the odd asparagus growing alongside the river trails to be harvested by urban foragers and the occasional reference in historical reminiscences. Where were these gardens? Who were the gardeners? And where did they go? Can parts of the invisible be made visible again?

According to Cashman, there were "about 15 [Chinese gardens] dotted here and there in Edmonton" in 1935. The oldest of them, she wrote, was "on the south side flats near the low-level bridge." It was a story about a site near the Fifth Street (or Walterdale) bridge that inadvertently led to this author's efforts to learn more about Edmonton's early Chinese gardeners. A brief one-time meeting with a second-generation Chinese Edmontonian alerted me to the possibility that a Chinese immigrant had grown vegetables near the river where John Walter's lumber business stood prior to the devastating 1915 flood. This same gardener, my source noted, had later moved his business to a location on the Fort Road, where it would not be subject to flooding.

The first listing in Henderson's Directories of a Chinese gardener on the former site of John Walter's lumber business, 9226–107th Street, was in 1923. Chin Lock gardened there for two years before being replaced by John Anderson in 1925. Three years later, in 1928, Chin Lock is again listed as a market gardener, having relocated to the Fort Road, where he stayed and ran a market garden for the next fourteen years. When Chin Lock died, on September 1, 1942, at the age of sixty-one, the obituary in the *Journal* made no mention of his occupation and, observing Canadian practice, reversed the order of his names so that the family name Chin appeared after his given name, Lock.[6]

Chin Lock was not the last Chinese to garden on the John Walter site. Jung Suey is listed as having a market garden there from 1932 through 1936. Gee Gut took over the property and ran the market garden from 1936 through 1950. The 1950 Henderson's Directory lists the property as being vacant and, in 1953, fifty-two acres of riverfront land were turned over by the City of Edmonton to the Kinsmen Club to be developed as a park, a project that was expanded a decade later, in 1963, when the Kinsmen Field House was opened on adjacent land. Cancellation of Gee Gut's lease and the "sale" of land to the Kinsmen

Club in June 1953 combined "to render life [in Walterdale] uncertain for the surviving old-timers," wrote a former resident whose memoir of the area exists only as a rough manuscript in pencil.[7]

Hong Lee, whose garden so impressed Cashman in 1935, appears to have had one of the most established Chinese-operated market garden businesses of his generation. From 1923 through 1948, Hong Lee's garden was visible from the top of Victoria Hill. Its address was listed in Henderson's Directories simply as "Groat Flats" or "Groat Ravine," although its precise location on the flats may have changed from time to time. According to Cashman, its first location was east of the one she wrote about in 1935, "but when the municipal links were extended it picked up its beds and walked before the imperative demands of another nine holes." Hong Lee's garden was forced to move even further west in the 1940s when Victoria Golf Course expanded from an eighteen-hole to a twenty-seven-hole course, but, eventually, the building of the Groat Bridge, begun in 1949, put an end both to Hong Lee's garden and to the third nine holes of Victoria Golf Course.

Hong Lee was always listed in Henderson's Directories as "Hong Lee Co." and in this he was unique. One would like to conclude that this implied something about the company's size and the scale of the garden's operations, but this may be going too far. Cashman did note that the garden's products were marketed to produce firms and to city grocery stores and, while she did not exclude forms of direct sales to customers, either door to door or through the city market, neither did she mention them.

An aerial photograph of Victoria Golf Course, said to have been taken in 1945, clearly shows Hong Lee's garden and three small clusters of buildings that appear to belong to it. If Hong Lee himself occupied one, it is possible that Seto Deep, listed in 1927 as the manager of "Hong Lee Co., market garden," occupied another.[8]

A "particularly large gardening operation" run by Chinese and located in Riverdale was described by Allan Shute and Margaret Fortier in *Riverdale: From Fraser Flats to Edmonton Oasis*. Relying also on Henderson's Directories as their source, Shute and Fortier attempted to trace a

Aerial photo of Victoria Golf Course, Edmonton, c. 1945. Hong Lee's garden is visible. [Photo courtesy of the author]

lineage of owners from Hop Woo, in 1935, to Wong Hop in 1949, George Hop in 1952, and Hop Wo in 1956. At least one Riverdale resident identified the gardener as Yet You. But Henderson's Directories was not always accurate or complete and confusion over how to spell and record Chinese names further compromises the written record.

The garden Shute and Fortier describe, known in its neighbourhood as "The Chinaman's Garden," was located on leased property that had once been part of the J.B. Little brickyard. According to their account, when Little sold a portion of his land for a school, the Chinese gardener installed gates in the fences around his garden so that the area's children did not have to detour on their way to school. The "Chinaman," despite being portrayed as a benign presence in the neighbourhood, emerges from this history as faceless. The garden, however, is described

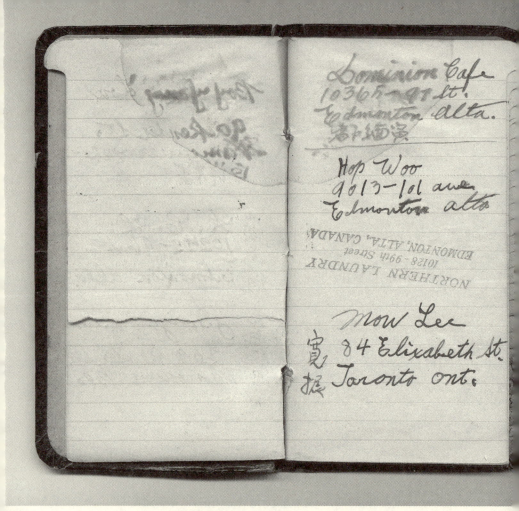

A page from Bark Ging Wong's black notebook, n.d. [Courtesy of Ging Wei Wong]

by a Riverdalian as a "delightful sight," consisting of "row upon row of lovely cabbages, radishes, lettuce, corn and carrots" that were "juicier and sweeter than any grown in our own gardens."[9]

It is the case, as Shute and Fortier note, that Hop Woo was first listed in Henderson's Directories as a gardener at 9012–101st Avenue in Riverdale in 1935. He continued to be so listed from 1936 through 1940, after which, until 1949, the occupant was simply described as "Chinese." Nowhere in Henderson's in the 1940s is Bark Ging Wong's name to be found, although it is certain that a man by this name worked and, at least for a time, lived there.

Young See Wong, Riverdale garden, Edmonton, 1949. On the back of the photo are the words "1949 Tak at Riverdal Mrs. B.G. Wong." [Photo courtesy of Ging Wei Wong]

In Bark Ging Wong's black notebook, the one in which he kept the names and addresses of contacts and possible employers, Hop Woo's name and Riverdale address are recorded. This same address is later recorded as his own on several documents issued to Bark Ging between 1940 and 1947, including the registration certificate issued in 1940 that he (and other Chinese in Canada) were obliged to carry and his 1947 application for Canadian citizenship. In 1949, the year Bark Ging's wife, Young See, was finally able to join her husband in Canada, a photo was taken of her in the Riverdale garden.[10]

The role played by Chinese market gardeners in their communities is not easy to decipher. On the one hand, the gardens seem to have been regarded favourably by neighbours, and the produce from them, especially when peddled door to door, was purchased and appreciated. On the other hand, the gardeners were rarely known by their neighbours, except as casual acquaintances.[11] In the early years, Chinese gardened on leased properties, making their tenure insecure and emphasizing the fact that they were not permanent members of the communities they inhabited. They were single men whose wives were not, until after 1947, allowed to join them. They had no children to assist in the process of integration. Apparent discrimination, such as occurred in 1927 when a Chinese market gardener was refused a stall at the city market on account of his ethnicity, was not uncommon.[12] And, even when there was goodwill on both sides, language was often a barrier. So, when a nameless gardener living at 9012–101st Avenue during the 1940s acquires a name, a family, a past, friends, and a future as a market gardener in Edmonton, a light is cast on at least one corner of the overall picture created by Edmonton's early Chinese gardeners.

In November 1975, long after their oldest son Kwan had completed his PHD at the University of Alberta in plant physiology and biochemistry and settled out of province, and two months after their youngest son Wei left for Ottawa to train for a job with Transport Canada, Bark Ging Wong and his wife, Young See, stopped market gardening and retired. In 1979, they moved to a house with a kitchen garden in the Dickinsfield area of the city. Bark Ging and Young See had, against considerable odds, earned their living in Edmonton as market gardeners and had raised three sons on their earnings. Wong's story, shadowy as parts of it remain even to his sons Kwan and Wei, casts some light on an aspect of Edmonton's gardening history that has faded from public memory.

Wong was born on July 15, 1908, in Chew Ging, in the district of Sin Ning, province of Kwungtung (today Guangdong), China. His father, Lip Sain, was the only son in a family of five. Bark Ging, third child but first male child in a family of seven children, would have had special

Bark Ging Wong's immigration certificate, August 20, 1921. [Courtesy of Ging Wei Wong]

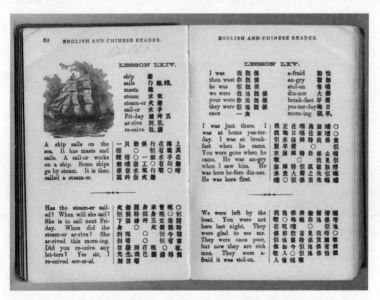

A page from Bark Ging Wong's Chinese–English reader. [Courtesy of Ging Wei Wong]

responsibilities within his family. It is undoubtedly these responsibilities that lay behind his arrival in Vancouver on the *Empress of Asia* on August 8, 1921, just three weeks after his thirteenth birthday. The look on the boy's face, captured in the photo on his immigration certificate, is not child-like. Resolve, stoicism, dignity, and fear can all be read into his expression. Whatever he might have felt and thought, however, his first task would have been to establish himself and find work.

Bark Ging was sponsored to come to Canada by a benefactor who lived in Vancouver. No head tax was paid on his behalf, a sign that his sponsor was either a merchant or a scholar and thus exempted from the tax. The sponsor's son, Alfred Wong, later became a close friend of the Bark Ging Wong family and was referred to within the family as "Uncle Alfred." It is likely that Bark Ging was taken into this family in Vancouver until he learned how to negotiate for himself in his new country and could earn his own living.

Bark Ging always kept an English–Chinese dictionary, a phrase book, and a reader close at hand and he referred to them often. He may have acquired these soon after his arrival in Canada. According to a story he told his children, he and some others were taught English by a "kindly gentleman" in Vancouver. Although he never became proficient in the language, Bark Ging learned to communicate in English. He was literate and informed, a regular reader of Chinese newspapers. He kept accounts for his business and, when he died, left behind two black notebooks, one with the names of contacts across Canada and the other with long lists of menu items, written in English with Chinese translations and notes.

The black notebooks, along with various official documents and casual bits of information passed on to his children, provide clues as to the course Bark Ging's life might have taken between his arrival in Canada in 1921 and his return to China in 1930 to visit with family and to marry. It would seem, according to these, that he left British Columbia for Alberta in 1924, at age sixteen, working his way slowly east and north to Edmonton. Kwan remembers his father speaking about the towns of Revelstoke and Golden, British Columbia, for example, and Wei still has an inscribed gift box containing a razor that was given to Bark Ging in the town of Retlaw in southern Alberta by Gum Lang, possibly his employer. Bark Ging also spoke of having worked in a mining camp, almost certainly in the camp kitchen, at Lake Wabamun, just west of Edmonton. It would seem likely that he began to work in the Riverdale market garden with Hop Woo sometime after his 1930 marriage in China to Young See.

Bark Ging made an extended visit to China in the fall of 1930. He returned to Edmonton in May 1931, a choice of dates that is consistent with the life of a market gardener. In addition to meeting and marrying Young See, Bark Ging would have met his youngest sister, born in 1924 and the seventh of Lip Sain's children. This same sister later immigrated to the United States and was reunited with her older brother in Edmonton half a century after their first meeting. The 1930 visit was the last time Bark Ging saw his parents.

A major marker in Bark Ging Wong's life was his marriage to Young See on December 30, 1930, a marriage almost certainly arranged by the two families. Both partners were young, Bark Ging not quite twenty-three years old and Young See four years younger. When Bark Ging left in May to return to Canada, Young See stayed with her husband's family in Chew Ging, where she would have been expected to serve and care for her in-laws until such time as she could join her husband. As it turned out, eighteen years were to pass before the couple could begin their married life together in Edmonton. For Young See, and the family into which she had married, it was an eventful period marked by war, displacement to Hong Kong, the death of Bark Ging's parents, and the adoption of a son. For Bark Ging, who contributed financially to the well-being of his extended family in China, it meant long, hard hours of work and the hope that, eventually, Young See and their son would be able to join him.

Bark Ging Wong's decision to become a market gardener rather than to work in a laundry, restaurant, or grocery store, other options open to Chinese at the time, was one he rarely spoke about. He may have had some gardening experience as a very young boy; Guangdong was, after all, a poor and primarily agricultural province at the turn of the twentieth century. Restaurant hours, which are long and irregular, were certainly incompatible with the family life he aspired to. He despised laundry work, which he had done briefly in Vancouver as a new immigrant. However, the explanation he gave his eldest son for choosing to market garden for a living was based neither on practical considerations nor on personal preferences but rather on principle. According to his eldest son, Kwan, Bark Ging loathed the idea of service, of having to jump to anyone's commands other than his own. What he liked about Canada, and his main reason for preferring Canada to China, was the freedom it gave him to be self-determining. He chose gardening, with all its risks and hard physical demands, partly because he enjoyed it, but primarily because he could be his own boss and the equal of any other Canadian.

It is likely that Bark Ging Wong made his first connection with the Riverdale garden sometime between 1935 and 1940, the years during which Hop Woo was listed in Henderson's Directory as the occupant there. Hop Woo was not known to Bark Ging's children, except as a name in a black notebook, although they do know that sometime in the late 1940s their father entered into a partnership with Chew Lung, the Riverdale gardener who was known in his neighbourhood as the "Chinaman," George, to buy some property in the area of the city known as Calder. The business relationship between Bark Ging and Chew Lung would explain why Young See was pictured working in Riverdale shortly after her arrival in Edmonton in 1949.

The partnership between Bark Ging Wong and Chew Lung, which ended in 1958, was a somewhat uneasy one; the two men often disagreed on issues that affected them both. For instance, Chew Lung was firm in his refusal to replace their two workhorses with a tractor, despite the logistical difficulties of moving the animals back and forth between Riverdale and Calder. One winter, while the horses were stabled on a farm outside the city, one of them was killed by a wild animal. Bark Ging decided that the time had come to modernize and, in a move that anticipated the end of the partnership, he purchased a tractor from his own savings. The two men were again forced to work out their disagreements when the time came to sell the Calder site, which happened after several wet years during which the property did not produce. Chew Lung was willing to accept the low offer for the property that was made to them by a somewhat shady real estate agent, while Bark Ging, who sensed that the agent was preying on them because they were non-English speakers, resisted the offer. Finally, through a complicated transaction and with the help of a lawyer, Bark Ging was able to get something closer to a fair price. From this point on, he worked only for himself.

If the partnership with Chew Lung was problematic for Wong, an earlier relationship, formed with the provincial gardener E.J. (Ernie) Stowe, was the opposite. Exactly how, or when, Wong came to meet Stowe is not clear, but the relationship appears to have been based on mutual respect and affection and it proved to be both morally and materially important to a Chinese gardener whose goal was to become productive and independent, as well as to have his family around him. Stowe, who began to work for the provincial government in 1913, had risen to become the chief provincial gardener for the Province of Alberta and was living in a cottage behind Government House in the 1940s when he must have met Wong. Perhaps Stowe had hired Wong at some point to work on a particular project, but he was most certainly instrumental in obtaining permission for Wong to garden a plot of land on the riverbank located immediately below Government House. In 1947, when Wong filled out a declaration of his intention to become a Canadian citizen, naming the Riverdale site as his address, he was described as a self-employed gardener "at the rear of old Government House." According to Kwan, who arrived in Edmonton the year before the provincial gardener retired from his position and moved to Victoria, British Columbia,[13] Stowe made city water available to Bark Ging for the plot below Government House, thus giving him an edge on the timing of his crops. And, as a testament to the strength of their personal relationship, Kwan understood from his father that Stowe had assisted in some way with the arrangements to bring Young

Studio photo of Young See and Bark Ging Wong, Edmonton, c. 1950.
[Photo courtesy of Ging Wei Wong]

See to Edmonton from Hong Kong in 1949, perhaps by helping with the application process. Even after Young See arrived, when the couple was first living together on the Calder site, Stowe remained a friend of the family.

When Kwan first arrived in Edmonton, he had a chance to meet Stowe and form an impression of the strength of the relationship between the provincial gardener and his father. The Wong family had been invited to the Stowes' house for a Christmas lunch in December 1951. Conversation may have been somewhat limited because Young See did not speak English and Kwan was very newly arrived in Canada. But Kwan remembers Stowe as a somewhat fatherly and down-to-earth person, who, in addition to being a gardener, kept bees and produced honey. Before the family left, Stowe presented Young See with several jars of honey, a commodity she could not have purchased and one that she valued partly for its known health benefits and partly for the gesture of friendship it implied.

The arrival of Young See in the spring of 1949 opened the way for a more settled life for Bark Ging and one that, for the first time, could be constructed around family. Shortly after her arrival, the couple must have decided to have a formal photo-portrait taken, a photo that marks the beginning of their life together in Edmonton.

Young See, born September 10, 1912, celebrated her thirty-seventh birthday shortly after coming to Canada. Unlike her husband, she never learned to speak English, although she was sociable and made many friends. She spoke little about life prior to her arrival in Canada, leaving her children with no account of her upbringing and no information about their maternal grandparents. A photograph, possibly taken to send to Bark Ging in Canada prior to their marriage, is the only image remaining from her early years.

Young See did speak a little to her children about the years she spent as part of her husband's extended family. Her job as daughter-in-law, she told them, was to care for her husband's parents. She had to learn how to please them, how to answer her father-in-law, and how to cook rice exactly as they wanted it cooked. Her status, according to her son Wei, would have been closer to that of a servant than it was to that of a family member. Bark Ging's mother died in 1939 and his father in 1941, releasing Young See from her duty to care for them but leaving undiminished her role as an integral part of the extended family.

On May 25, 1950, less than a year after her arrival in Canada, Young See gave birth to Lilly Mee Wong. On June 29, barely a month later, the baby died. Young See never spoke about Lilly Mee's birth or death. Not until shortly before their mother's death, in 1990, did Young See's children find their sister's gravesite. One of Young See's last expressed wishes was that a proper gravestone be erected for her daughter.

Photo of the Wong family, Edmonton, 1956. Left to right: Kwan, Bark Ging, Wei, Young See, Wayne, and Joe. [Photo courtesy of Ging Wei Wong]

The hole created by Lilly Mee's death was not long in filling. Fook Kwan, adopted in China during the years before Young See's arrival in Canada, arrived in Edmonton in 1951 at the age of eleven. In October 1952, Young See gave birth to a son, Ging Wayne; in January 1954, Ging Wei was born; and in January 1955, Bark Ging's nephew Joe Wong arrived in Edmonton just in time for Wei's first birthday, having travelled from Hong Kong through Vancouver to join the family and to work with Bark Ging at market gardening. Thus, within the space of three years, Bark Ging and Young See found themselves with a family of four boys.

The family's first house, 12782–113th Street, was in Calder (now Lauderdale), on the northwest corner of an eight-square-block property now known as Grand Trunk Park.[14] This is the address on Lilly Mee's birth certificate and was the first listing for Bark Ging and Young See in Henderson's Directory. It was the site from which the family market gardening business began.

The Calder house was small for a family of six—two bedrooms, a living room, and a kitchen, all of which were on one floor. The kitchen was equipped with a wood stove. There was no indoor plumbing and no telephone. Drinking water was trucked in. In addition to gardening on his own property, Bark Ging planted cabbages on three acres to the north, on property owned by neighbours, the Ewings.

On September 1, 1958, before Wei and Wayne began to attend school, the Wong family moved to a three-bedroom bungalow in the Dovercourt area of the city, a house that was miles away from the acreage Bark Ging was to buy for his market garden. Unlike the Calder house, the Dovercourt house had telephone, power, water, and natural gas. There was a large backyard that included a kitchen garden, a hotbed used to start seedlings for the market garden, and a single detached garage. It was close to schools for the children: Dovercourt Elementary, Sherbrooke Junior High, and Ross Sheppard High School. By the time it was sold, in February 1979, Joe had long left gardening behind to enter the grocery business, while Kwan and Wei were settled in their careers and living away from Edmonton. Bark Ging and Young See, who lost their son Wayne in 1978, were on their own.

The last years during which the Wong family lived in Calder were particularly wet ones, rendering the Calder property close to unusable for gardening. In order to earn a living, Bark Ging turned his hand to managing a larger gardening enterprise on the east end of the city. The property, located on the northeast corner of the Clover Bar Bridge, just opposite present-day Rundle Park, bordered the North Saskatchewan River. A former coal mining site, the land was bought by a group of Chinese whose plans to create a market garden there never materialized. While the upper part of the property was converted to a gravel pit, in 1956, Bark Ging leased the lower part, approximately twenty-five acres according to Kwan, from Harry Yee and Mah Foo, a rendering of events corroborated by longtime friends of Bark Ging and Young See, Bob and Flora Chow. One of the five empty bunkhouses on the site was refitted to provide cooking and eating facilities for the gardeners, while another one was retained as a bunkhouse. The barn was used to store equipment. It was while planning this venture, according to Kwan, that Bark Ging realized he would need reliable help and began to make the arrangements to bring his nephew to Canada.

All members of the Wong family participated to make the Clover Bar garden venture a success. While Bark Ging and Joe worked there full-time during gardening season, the other family members worked whenever they could. As soon as Kwan finished school, he accompanied his mother and two younger brothers on a bus ride to the end of town where Bark Ging picked them up. During school holidays, the entire family drove together to Clover Bar. While the rest of the family worked, the two younger Wong children amused themselves, watched over by the farm cook, who, in deference to his grandfatherly status, was referred to by the younger children as Charlie *gung*. When the meal was ready, according to Wei, Charlie *gung* would summon the workers by raising a white flag on a pole at the side of the cookhouse.

Income tax records for the period during which Bark Ging leased land at Clover Bar show that in 1956 he paid three hundred dollars in rent for this property and fifteen hundred dollars in wages. Between 1957 and 1959, the annual rent rose to five hundred dollars and Bark Ging paid amounts between $2,070 and $4,020 in wages. The records suggest that Bark Ging ran the Beverly operation on his own, a scale of operation he was not to continue on the Wongs' next property at Namao.

In October 1959, Bark Ging and Young See purchased two parcels of land, approximately ten acres, at 9404–157th Avenue, then in the Sturgeon subdivision of Waldemere. The Wong family always referred to the acreage as "the Namao farm," probably because of its proximity to the planes flying in and out of the air base. Even while weeding under the hot sun, Wei remembered, it was possible to get a good view of the Namao Air Show from it.

The farm was essentially rural during the years the Wong family owned it. In addition to country houses and recreational acreages, neighbours included other market gardeners. Immediately to the south of the Wong property, for example, was a market garden run by Cheung Gee, son of Gee Gut, the last person to have gardened at the Walterdale site. In 1971, the Namao farm was annexed by the City of Edmonton. In 1975, the year the farm was sold, a property assessment described the site as "future subdivision land," with market gardening named as an interim use.

There were four small structures on the Namao farm. The largest of these was the transposed Calder house that was moved there and placed on a cement foundation to be used for storage purposes. No thought was given to hooking this house up to services, so the farm was operated without power, gas, electricity, telephone, or plumbing. Drinking water, and occasionally water for germinating seedlings, was brought from Dovercourt, and an outhouse was the alternative to indoor plumbing.

Three shacks that came with the property were put to use. One stored forty-five-gallon drums of purple gasoline, motor oil, and lubricants that were delivered by the United Farmers of Alberta co-operative. A second shack housed the Ford 9N series tractor, wheel hoes, wheelbarrows, pesticide applicator, pesticides, and farm clothes such as coats and rain gear. A smaller shack nearby had a pot-belly stove vented outside and, to the children's delight, came stocked with old English language books, newspapers, comics, and magazines.

From 1959 through 1975, the family commuted daily between early May and late October to grow and sell vegetables. All members of the family were expected to participate; the proceeds were the family's sole source of income. The daily routine during gardening season revolved around farm work, with the children participating part-time when they were in school and full-time in July and August.

Mornings began with preparation for the day ahead. While Young See made breakfast and prepared food for lunch, Bark Ging delivered vegetables that had been loaded into the truck the night before.

When the deliveries had been completed, Bark Ging returned to pick up Young See and, on nonschool days, the children, for the drive to the farm. Once there, if a large order needed to be filled the next day, the morning's work began with harvesting; otherwise, it began with weeding and hoeing, leaving harvest for the afternoon.

Lunch required a commute back to Dovercourt, where the family ate the food Young See had prepared early that morning. Bark Ging took time for a smoke and a glimpse at the Chinese newspapers. Young See organized rice, soup, and other dishes they took to Joe on their way back to the farm, stopping to look after the grocery store while he ate. On the way back to Dovercourt from the farm, the family again stopped at Joe's, this time to pick up the cooking pots, a stop that gave Joe a chance to ply his young cousins with chips, pop, and chocolate bars. At the end of the day, Bark Ging and Young See prepared the evening meal together, a meal that was often eaten late during the gardening season.

The first year at Namao, in preparation for planting, Bark Ging burned off the ten acres before plowing. Every spring thereafter he plowed with the aid of his old grey Ford tractor. In the fall, if there was time before the snow fell, he would take the same tractor around the acreage with the disc to break up the vegetable matter. The Ford tractor, already more than fifteen years old in 1959, was kept in perfect running order by Ed Reinert, a family friend and mechanic whose tuning and repair work were generally recompensed by a home-cooked meal with abundant conversation and plenty to drink. The farm was sold in November 1975 at the end of the last harvest season. The old grey tractor, still in running order, was sold for a mere six hundred dollars, a bargain Wei thought for the antique it must have been.

Planting was carried out in stages. Some crops were started from seed in hotbeds behind the Dovercourt house; others, mainly cabbages and cauliflower, were started in hotbeds at the farm; yet others were seeded directly into the ground. Radishes, the earliest crop to be ready for market, were simply broadcast by hand over a large area.

Many vegetables, including the Chinese varieties, were sowed in rows with the Planet hand-pushed seeder, a job Bark Ging always carried out himself. String was used to line up and measure the rows, leaving enough room between each one for the wheel hoe that would be used for weeding throughout the season. Cabbages and cauliflower were transplanted manually as seedlings after being started in hotbeds. The hotbeds were fashioned by Bark Ging from salvaged windowpanes, which could be easily moved around to allow for ventilation and to provide either shelter or extra heat as required.

Seeds came from several sources. Most were bought from local seed companies, mainly Pike and McKenzie Seeds. 'Golden Acre' cabbage, described in the 1941 Pike seed catalogue as a good variety for market garden purposes, was a major crop. Other vegetables planted included peas, beets ('Early Wonder'), spinach ('King of Denmark'), Spanish onions, pickling cucumbers, carrots, turnips, parsnips, cauliflower, squash (winter varieties and zucchini), green and yellow beans, celery, tomatoes, radishes, corn, parsley, and dill weed.

Cauliflower, a staple, was a tricky vegetable crop to grow. Every year, to prevent the heads from yellowing in the sun, the Wongs tied each plant with raffia when the heads began to form. Bark Ging's notebook lists several varieties of cauliflower, including 'Early Erfurt', 'Sluis Brothers', 'Eureka Early', 'Snowball', 'Danish Giant', and 'Dry Weather', suggesting either that he experimented with this vegetable, or that, in order to be assured of a crop, he had to be prepared for a wide variety of weather possibilities.

Seeds for Chinese vegetables such as bok choy, suey choy, gai choy, gai lan, lo bok, Chinese snow peas, and Chinese mustard were more difficult to acquire. Initially, Bark Ging bought seeds by mail order from Chinese shops in Vancouver. In Edmonton, he obtained seeds from friends, mainly women who saved them from year to year, or from Chinese shops in Chinatown. After a trip to Toronto, Bark Ging returned with some special seeds he had acquired from a friend who farmed—bok choy and a particular variety of gai choy that formed a seed head. Most importantly, Bark Ging and Young See produced their own seed for Chinese vegetables by selecting them from the most healthy specimens in the garden.

Good plant specimens of suey choy, bok choy, and gai choy were identified and staked. When they had matured and produced seed pods, they were cut off at the stem and dried, in the sun if weather permitted or by hanging them in one of the farm buildings. Once dry, the seed pods were stripped into a bucket or flailed against the side of a big container before the stems were discarded. On calm sunny days, the mix of tiny seeds and seed pods was put into a twenty-to-twenty-four-inch circular bamboo tray made of flat, woven wicker with a one-inch side. In the Wongs' Toisan dialect this tray was referred to as a *boo kee*. Partial and empty seed pods were discarded by hand. Then, holding the tray and giving it a deft flick of the wrists, Bark Ging or Young See would launch the mix into the air so that the breeze could blow the fragmented seed pods away, leaving the thousands of tiny seeds to fall back into the tray. Some of these were given to Chinese friends and relatives. The rest would sustain the farm's Chinese vegetable supply for the following season.

The farm came with some hardy fruits which, though they were never sold, were sometimes harvested for the family. There was the inevitable stand of rhubarb. At the far north end of the field, there were some raspberry canes. The southeast part of the property had some saskatoon trees that bore berries. And, when Kwan was studying botany at the University of Alberta, he discovered some gooseberry bushes about midway up the fence on the east side, bushes that he said were a native berry that yielded small, inedible fruits.

Cultivation was done almost entirely by hand with the aid of garden tools and implements, some of which were handmade out of recycled materials. Watering, for example, was done only when required and then with homemade implements. Used forty-five-gallon oil drums, scavenged or bought, were cut out at the top and set up beside the sheds to collect rainwater. When rain was scarce, the same drums, these ones fitted with a tap near the base and a short hose, were used to transport water from Dovercourt. To get the water from the drums to the fields, buckets were fashioned from five-gallon metal containers with the tops cut off and a wire attached on opposite sides. These were carried, two at a time, balanced on the shoulders at either end of a five-foot pole made from the stripped branch of a tree. The watering implement was a small dipper made from an Export tobacco tin (Bark Ging's brand for rolled cigarettes) that was fastened to a stick. Every seedling transplanted was carefully watered in. Then, to prevent evaporation of the precious water on hot days, a thin layer of soil was sprinkled manually over the moist soil. A variant on the Export dipper was the tool used for watering seeds in the hotbeds; it was a large can that had many nail holes punched into the bottom before being attached to a stick.

Most weeding was done by hand using old kitchen knives or purchased stirrup hoes. Hand-pushed wheel hoes, when they could be used, were more appealing to the children, who, when out of sight of their parents, raced them between the rows, stopping only to attempt to hide the occasional casualty.

Elephant brand fertilizer, bought in fifty-pound paper sacks, was applied to the vegetables in three ways. First, it was applied sparingly by hand to the hotbeds before watering in the seeds. A second alternative saw Bark Ging putting a couple of bags of fertilizer on the back of the pickup for a slow drive through the crops, stopping strategically while members of the family filled up large empty tins with fertilizer and delivered a small amount to each plant. The third way was for Bark Ging to put it in the Planet seeder and deliver it adjacent to the established rows of vegetables.

Insect and animal pests were dealt with only when they became a particular problem; cutworms were a persistent enemy. Aldrin, promoted in seed catalogues at the time but now banned, was used against cabbage worms, applied with a hand-held applicator that was piston-driven and operated somewhat like a bicycle pump. Gophers, the only serious animal pest contended with at the Namao farm, were trapped and then "finished off" with a .22 calibre rifle kept at the farm for the purpose.

Two machines were vital to operations at the farm. The old Ford tractor, with attachments such as a two-bladed plow, a harrow, and a wooden skid used to haul people and tools into the fields and bring produce back, remained in service from 1959 to 1975. The farm truck, on the other hand, was replaced several times and was always maintained by another friend of Bark Ging's, Henry Beech. The last truck, Wei remembered, was a blue Ford F-250 three-quarter-ton pickup, purchased new from Waterloo Ford Mercury in 1972 for $3,400. Between 1972 and 1975, while Wei was attending classes at the University of Alberta, he frequently drove it early in the morning, Bark Ging in the passenger seat, to deliver produce to wholesalers before taking the bus from Dovercourt to the university campus.

Bark Ging never sold directly to consumers, either by peddling vegetables door to door or by selling at the Edmonton City Market. He preferred to sell in bulk, either to wholesalers or to independent businesses. Woodwards at Westmount, Macdonalds Consolidated, Scott National, and the Brown Fruit Company were among his wholesale customers. Chinese restaurants and stores such as the Kwong Hing Company on 97th Street were also reliable sales outlets.

Bark Ging wrote all of the produce orders in English in a simple order book, the top copy of which would serve as the customer's invoice. Because wholesalers paid their suppliers according to the market prices for the day, amounts owing were left for the buyer to calculate. For example, an order form presented to Dominion Fruit in September 1975 records only the amounts delivered—one thousand pounds of suey choy; twenty pounds of gai lan; forty-five pounds of choy sum; fifty-six, fifty-pound bags of cabbage; eight fifty-pound bags of beets; and 510 pounds of bok choy. Payment from these customers was generally mailed and would arrive weeks after the invoice date. Orders to Chinese establishments, on the other hand, show the price tallied, and most likely collected, immediately upon delivery. On

October 21, 1975, the Kwong Hing Company bought from Bark Ging sixty-five pounds of bok choy at thirty-five cents a pound and twenty-five pounds of gai choy at thirty cents a pound. Bark Ging's last season of operation, 1975, was a good year. Records show, for example, that in a period of slightly less than a month that year he sold 419 fifty-pound bags of cabbage.

Harvest was one of the busiest times of the gardening year and ample preparations for it were made ahead of time. Throughout the winter and early spring, wooden crates, delivery boxes, and bushel baskets were collected by Bark Ging to be used at harvest time. He even collected scrap wood for repairs. It was the children's job to remove nails and damaged boards from produce crates and then to assist in repairing them for re-use in the summer.

The Wongs did not have a root cellar for vegetable storage, so harvesting was always done in response to orders. In the late fall, if snow came early, the family would try to harvest the entire crop of cabbages, bag them in fifty-pound sacks, and store them until they were sold.

Green cabbage was the Wongs' main crop and help from one or two close friends was sometimes needed to harvest it. Cold weather often made it a tough job, but even if the children were numb with cold they knew better than to complain. Complaining was not tolerated. On the other hand, they were quite capable of turning the situation into fun for themselves. As harvest progressed, they tossed around small cabbages like footballs, ran through the rows of mature dill weed, and kicked up mounds of spoiled or rotting suey choy.

Harvested vegetables were brought to the south entrance of the farm to be cleaned, weighed, prepared for delivery, and loaded onto the truck. Two metal tubs, each one the size of a bathtub, were set up between the tractor garage and the smaller shack to wash the root crops. Root vegetables were soaked and scrubbed clean in the first tub and then rinsed in the second tub before being packed into crates or sacks. Parsley was dunked into the rain barrels to keep it fresh and out of the sun. At the end of the day, the bunches were counted and packed into crates.

Cabbages were collected in the fields in bushel baskets before being carried to the central work area, where they were weighed and packed into fifty-pound sacks. The sacks, some brown and some green, were collected year-round by Bark Ging, who paid a nominal price to retrieve them from wholesalers. After being filled and weighed, the sacks were sewn up with large sewing needles, threaded with bailing twine, and tagged with a label that identified the contents and their origin. These tags, purchased by Bark Ging during the winter, occupied the children on many an afternoon in winter or early spring. Using a rubber stamp and purple ink pad, they imprinted them with the words "B.G. Wong, 12261-134 St, Edmonton, Alberta," after which they carefully hand-printed the words "Canada #1 Cabbage."

Radishes, cucumbers, squash, and zucchini were also harvested into bushel baskets. Raffia, cut into appropriate lengths with one chop of the cleaver, was used to tie bunches of radishes, green onions, carrots, and beets before they were packed into crates for delivery.

At the end of each day, the truck was loaded according to the following day's delivery schedule. Cool water was sometimes splashed over the crates just before loading to keep the contents fresh. A burlap quilt made from potato sacks was thrown over the pickup box and tucked in at all sides before the tailgate was closed. If necessary, the family made a quick stop at the farm gate to check the oil and pump gas into the truck from a five-gallon metal jerry can. Finally, after all equipment had been stored, and the shacks had been padlocked, it was time to close the main gate and return home for dinner.

The Wong family lived entirely from their market garden income: neither parent took on additional paid employment during the winter months. That income, as recorded in twenty years of tax returns (1956–1976), varied from a high of just over six thousand dollars to a low of just under six hundred, figures that seem astoundingly low by contemporary standards, despite the fact that vehicles and gas were written off as farm expenses. The winter's budget was based on the profits calculated at the end of each summer and, according to Wei, the family never lacked for food or other basics. In fact, long after the growing season was over, Bark Ging went through his earnings and bank books and wrote a cheque to each of his children for their help during the past season, encouraging them to save it in the bank or to buy Canada Savings Bonds for use in their future education. Frugality, a response to economic realities, became a way of life that informed family activities without restricting them.

Family life in Edmonton for Bark Ging and Young See involved balancing the opportunities presented by a new country with a desire and an instinct to preserve the values and traditions from their Chinese heritage. Integration into Canadian society was not so much a goal as a necessity. Discrimination, while rarely mentioned by either parent to the children, must certainly have been encountered, particularly by Bark Ging in the years before the 1947 repeal of the Chinese Exclusion Act. Nevertheless, while the majority of the Wong's socializing was done within the Chinese community, English-speaking acquaintances, many of them made through work, often became family friends.

Language was a barrier between the Wongs and Canadian society, especially for Young See. Bark Ging could communicate with English-speaking friends, but his English vocabulary was limited and it is likely that he translated his thoughts from Chinese. While Bark Ging and Young See wanted their sons to excel at school, they also wanted them to speak their Toisan dialect at home. It was something of a shock to the family when Wayne was denied entrance to Grade 1 because he did not speak English.

It is almost certain that Bark Ging and Young See thought of themselves as Chinese first and Canadians second. They always cooked traditional Chinese foods, kept Chinese traditions, and enjoyed Chinese celebrations, passing on the rituals and customs to their children. One such tradition was the occasional family excursion to the Dreamland Theatre downtown to attend a Cantonese opera. Cantonese opera gave Bark Ging and Young See an opportunity to relax with Chinese friends in a familiar language and culture. These exclusively Chinese entertainments did not, however, prevent them from responding favourably to any reflection of Chinese culture in the North American media, even

Reunion of Bark Ging Wong with his oldest sister in China, 1976. [Photo courtesy of Ging Wei Wong]

when a Chinese character was portrayed by a non-Chinese. Hop Sing on the television show *Bonanza* was a favourite, as was David Carradine playing Kwai Chang Caine in the *Kung Fu* series. They watched Martin Yan on the *Yan Can Cook* cooking show and were, according to their children, pleased and surprised to be approached in a supermarket aisle with a question about how to prepare one or another of the Chinese vegetables displayed.

Bark Ging and Young See relied on both Chinese and Canadian medicine to deal with health problems. Chinese herbal remedies were generally tried first; if these failed, Dr. Max Dolgoy, the family's general practitioner, was called. The Wongs always kept cinnamon sticks and other ingredients used to make tinctures and lotions. At the same time, Dr. Dolgoy became a trusted resource and friend, who was always invited, along with his wife, Rachael, to share major family events.

Wong Family, Edmonton, 1964. Left to right: Joe, Kwan, Wei, Young See, Wayne, and Bark Ging. [Photo courtesy of Ging Wei Wong]

Not until they retired did Bark Ging and Young See have the leisure and resources to return to China. A trip there in 1976 gave Bark Ging the opportunity to be reunited with his oldest sister, not seen since 1930. This reunion, captured in a photo, clearly identifies the siblings as having lived in different cultures. The return from China to Canada must have been bittersweet for Bark Ging and Young See. On the one hand, they were returning home to their children; on the other, they must have realized that they would not see family in China again.

Although Bark Ging was physically separated from his parents, siblings, aunts, and uncles at a young age, his sense of responsibility for that family and his connection to it never left him. Cut off as they were from family in China, and having undergone life experiences that demanded stoicism and emotional restraint, both Bark Ging and Young See found ways to demonstrate love for their family in Canada.

They treated their three sons and their nephew Joe with affection and respect while expecting them to work hard and to excel. Photos of the Wong family taken while they were living in Calder and Dovercourt, with all members of the family neatly turned out and looking confidently into the camera lens, convey the easiness of family members with one another and their consciousness that they are presenting themselves as a family to future viewers.

Photos taken during this same period also suggest that Bark Ging and Young See were not restricted by a narrow sense of family. Good friends appear in photos with the regularity of extended family; business contacts and neighbours are often included in group photos. Bark Ging and Young See had a wide range of friends and acquaintances that extended well beyond the Chinese community. Bark Ging especially was comfortable with his elders, people like George (Sek) Wong, for instance, who appeared in many of the Calder photos. The Wong children addressed George Wong as *Sek bak* or "older uncle." He amused and amazed them by performing somersaults, feats of physical prowess they did not associate with the elderly. He and other of Bark Ging's friends were present in Wong family life as though they were blood relations. In some way they stood in for the relatives still in China and Hong Kong. Bark Ging cultivated these quasi-family relationships but never ceased to communicate with family members in China.

Generosity and trust characterized Bark Ging and Young See's dealings with friends and neighbours. Combining thrift with thoughtfulness, they ensured that mature but unsold vegetables from the garden were given to neighbours and friends before they spoiled. And, while they spent little on themselves, both Bark Ging and Young See were generous within their means. Bark Ging was known to have given to someone he considered more in need than himself the "scrap" meat he had just obtained from the butcher for his own kitchen. Kwan remembers his father saying, in a half-joking manner, that if money he had loaned to others was all paid back, he would be a rich man. And Young See, her family later learned, loaned a friend two hundred dollars to buy a sewing machine at a time when she could ill afford it.

The daily routine of the Wong family eased somewhat when the harvest was complete. Bark Ging, always the first to arise in the morning, was easier to live with in the off-season than he was in the summer. Before the rest of the family was up, he relaxed in his favourite rocking chair with a pot of Chinese tea and a hand-rolled cigarette. After lunch he took time to read the Chinese newspaper, look at the *China Pictorial*, and, with the aid of the phrase book and reader, peruse the *Journal*. If he went downtown to do errands or meet friends, he went by bus, never leaving the house without a fedora. On occasion he went to Chinatown to meet friends and play mah jong, outings that frequently resulted in Young See asking one of her children to call and remind him to come home for supper. On hot days he enjoyed a Lethbridge Pilsner beer with supper and on special-occasion meals he drank straight rye. A good meal would sometimes put him into a storytelling mood that his children later wished they had taken more advantage of.

Together, Bark Ging and Young See passed quiet time playing various games of solitaire. After Joe's mother moved from China to Edmonton, the two families would sometimes play mah jong together, although Bark Ging did not like to lose and was known to leave the table in a huff when this happened. Neither parent ever attended parent–teacher interviews, although the children were always eager to show off the results attained on their report cards, especially as all three boys were honour students. Family outings on weekends might be to visit friends in or near the city, or, on occasion, to visit Storyland Valley Zoo.

The Wongs rarely went out for meals except to celebrate special occasions such as a wedding or the annual banquet put on by the Wongs' Benevolent Association. Such celebratory dinners were held in restaurants such as the Lychee Gardens or the New World Restaurant, both of which were in Chinatown. Once or twice a year, the family purchased a take-out order of Kentucky Fried Chicken for a treat. Bark Ging would chop each piece into several smaller pieces that could be picked up with chopsticks. Important dates on the Chinese calendar such as Chinese New Year, Ching Ming Celebration, Mid-Autumn Festival, and Dragon Boat Festival were celebrated with the appropriate foods that were either prepared at home or bought.

Cooking was central to the functioning of the Wong household. Both lunch and supper were cooked meals that Bark Ging and Young See often prepared together. Traditional Chinese dishes and menus were standard fare and included steamed rice, soup, stir-fried meat and vegetables, and sometimes fish. Occasionally, Western-style food was cooked, but more often Bark Ging experimented with combinations that made use of Canadian ingredients in Chinese-style dishes to produce tasty combinations.

Bark Ging was an excellent cook and a skilled butcher. When he was able to obtain live chickens, he butchered them himself in the basement so he could use all the parts, including the blood, in traditional dishes. Fresh blood, for example, was steamed to a lightly gelled consistency and served as a dish.

Despite the Wong's frugality, the range and quality of ingredients used in the cooking was impressive. Bark Ging and Young See sometimes received wild game and fish from relatives and acquaintances. Ducks, geese, moose, venison, white fish, jack fish, and pickerel were thus sometimes on the menu. The jack fish was often made with shredded carrots and celery into fishcakes and pan-fried. In the summer, there was an abundance of fresh vegetables. Bok choy was dried and gai choy was pickled for use in the winter. Especially during their early years in Dovercourt, the Wongs bought Chinese foodstuffs from Vancouver, including tins of fried dace, preserved black beans, rice flour, preserved salt fish, dry bean curd, preserved bean curd, long grain rice, dried shitake mushrooms, oyster sauce, soy sauce, tinned abalone, puffed fish maw, dried oysters, dried squid, dried shrimp, dried scallops, Chinese herbs, and Chinese medicine. These items may have been purchased from the Tai Hing Company in Vancouver when Bark Ging and Young See were on holiday there and delivered to Edmonton later. Bark Ging was a shrewd consumer, seeking out the best prices, especially for the rarer items.

The Wongs' kitchen garden gave precedence to Chinese vegetables. During the Dovercourt years, after the hotbed plants had been transplanted to the Namao farm, the kitchen garden was set out with bok choy, suey choy, green onions, gai lan, snow peas, beans, peas, and kohlrabi. An apple tree, a pear tree, and plum trees were planted in the yard. Even after retirement, Bark Ging and Young See planted a kitchen garden for their own use and enjoyed taking vegetables to friends and relatives.

Bark Ging loved flowers and was proud of those he grew in the front and back yards of all his houses. Hollyhocks, bleeding heart, and peonies were favourite perennials, partly on account of their hardiness. Every year, he planted giant and small dahlias, asters, stocks, snapdragons, and pansies. For several years, according to Wei and Kwan, bedding plants were delivered in flats each spring to the Wong home by Ted McNeil, Stowe's son-in-law, a legacy of Bark Ging's days as a market gardener behind Government House.

Family vacations were rare for the Wongs, although the entire family travelled more than once to Vancouver to visit "Uncle Alfred" and his family, a relationship begun in 1921 that lasted throughout Bark Ging's lifetime.

Day trips were more common family outings and a frequent destination for these was Leduc, where they visited Bark Ging and Young See's best friends, Sew Wong and his wife Tong. Sew Wong always called Bark Ging *tsai lo* or "little brother," and the children referred to their father's friend as *Sew bak*—a tacit acknowledgement of the closeness of the relationship. Sew Wong worked at the Union Cafe on the main street of Leduc, just a few doors away from the family's home. The closeness of these two families was further cemented in 1981, when Bark Ging and Young See's youngest son, Wei, married Diane, the youngest daughter of Sew and Tong.

Bark Ging Wong died in November 1988. Young See, who had always suffered from somewhat fragile health, died less than two years later, in September 1990. Two of their children, Lilly Mee and Ging Wayne, had predeceased them. Neither Kwan nor Wei, both of whom had worked hard in the garden as children, chose market gardening as a career and neither had been encouraged in this pursuit. On the contrary, Bark Ging and Young See's hard work and sacrifice was intended to release their children from hard manual labour and the economic insecurity associated with market gardening, intentions that fit well with both Kwan and Wei's own inclinations.

Neither was market gardening the career Eric Chen's parents, Laotian refugees whose southern Chinese origins had been lost to a series of moves in search of political and economic security, would have chosen for their only son. Born in Laos in 1966 with the name Hua Phetsavanh, Chen arrived in Edmonton on October 30, 1979, with his parents and two sisters after spending two years in a refugee camp in Thailand. Chen, who goes mainly by his English first name and who officially reverted to the family name of his ancestors when he reached adulthood, was thirteen years old when he and the other members of his family were accepted as Canadian refugees. Chance, not choice, lay behind the family's Edmonton destination.[15]

The Phetsavanhs moved to Westlock where the children were enrolled in school and where their parents found work providing janitorial services to local businesses, including the town's IGA (Independent Grocers) store. To ensure that the janitorial business prospered, the Phetsavanh children often worked nights after school to clean buildings with their parents. The family expectation, as with the Wong family, was that education and integration into Canadian society would launch each of the children into a profession such as law, teaching, medicine, or engineering.

Unlike the Wong children, however, Chen was not a keen student; gardening was his passion. While still in high school, he found a summer job working for a market gardener in a town about twenty kilometres from his home, a job that appealed to him in all ways except for the use of chemical fertilizers and pesticides, a practice he disliked, although he had no models at the time for an alternative practice. Unclear as to the path he wanted to follow after graduating from high school, he worked for a few years for local greenhouse businesses.

Then, influenced by a sister who was attending Grant MacEwan Community College (now MacEwan University), he enrolled in some arts courses, an experience he later recognized to have been a waste of time.

In 1987, Chen successfully pressured his parents into purchasing twenty acres of land in northeast Edmonton on which he hoped to create a market garden of his own. The soil was poor, but, with help from family members, it was made to produce. Thus it was that, a little more than a decade after Bark Ging and Young See Wong planted their last market garden and a year before Bark Ging's death, Chen's Market Garden came into production.

Eventually, by a somewhat circuitous route and in an effort to assuage parents who did not want their son to be condemned to a way of life his grandparents had abandoned when they left China, Chen enrolled at the University of Alberta from which, in 1995, he graduated with a Bachelor of Science in Agricultural Economics, a clearer sense of his direction in life, a determination to reclaim his family's original name, and, most importantly, partner Ruby Law. Ruby (Pui Ling) Law, a Hong Kong native who graduated from the University of Alberta in 1995 with a bachelor of commerce degree and the intention of returning to her large family in Hong Kong, was persuaded by Chen not only to marry and stay in Edmonton but to participate in the market gardening business. With help from family members, they moved the business in 2003 to sixty acres of productive land just east of St. Albert and two miles north of Edmonton.

Peas on Earth is an organic market gardening business that sells directly to customers, mainly through farmers' markets but also from a shop area located below the Chen family's living quarters. Its proximity to two cities is consistent with its owners' belief that there is a symbiotic relationship between healthy urban centres and the rural or semirural hinterland that supplies local food; by selling locally grown organic produce to their urban customers, Chen and Law actively promote this symbiosis. The receptivity of their customers to the Chinese vegetable component of their market stall surprised Chen and Law at first but soon resulted in adaptations to their planting and marketing plans. Personal contact with customers, especially at their year-round stall at the Old Strathcona Farmers' Market, has allowed them to respond to customer interests and preferences and to educate their customers on how to use some of the less common produce, a business model that has resulted in gradual but profound and ongoing change to what is grown and how it is marketed. Indeed, Chen believes that without the market stall, Peas on Earth could not have been developed as a viable business, one capable of supporting a family and fulfilling its owners' desire to make a meaningful contribution to the community in which they live.

It took Chen a long time to acknowledge fully that his passion for gardening could be a vocation. Parental and cultural attitudes towards market gardening as a way of life acted powerfully to dissuade him for years. In the end, choosing to become a market gardener went hand in hand with the decision to reclaim his Chinese heritage. He wanted to raise the status of gardening as a career within the Chinese community. Learning that he was not the first Edmontonian of Chinese ancestry to make a living as a market gardener, that he was part of a tradition, came as a happy surprise and added an important dimension to his sense of the cultural context in which he operated. Unlike Bark Ging Wong and other Chinese immigrants who came to Canada and to Edmonton before the end of the Second World War, Chen could say that gardening was his first choice of occupation, one around which he could raise a family and help to build a stronger and healthier community.

Eric Chen and Wei Wong, whose involvement with gardening in Edmonton has been entirely within a Canadian tradition, are drawn to the Chinese Garden in Louise McKinney Park as an expression of a culture that both is and is not their own. Their reaction parallels that of author Marty Chan. For Wong, whose father gardened in nearby Riverdale, it has a particular relevance, one he has acknowledged with two donations to the Chinese Garden Society, a donation in memory of his parents and another to mark the first wedding anniversary of his eldest daughter, Jasmine. For all three, it serves as a reminder of what was left behind when their parents and grandparents left China to create a new reality for their children.

George Ng, who knew nothing about traditional Chinese gardens when, in 2000, he agreed to join an as yet informally constituted group of people whose aim was to build such a garden in Edmonton, became a firm believer in the symbolic significance of the project he helped to realize. Similarly, architect Francis Ng knew little about traditional Chinese garden design when he agreed to participate in the project. More than a decade later, with the basic elements of the design in place and several of its key components completed, both men have become apologists for the garden and both have identified with its potential to symbolize Chinese participation in the ongoing project of city building.[16]

Alison Hardie, in her introduction to *The Chinese Garden*, by Maggie Keswick, suggests that "[o]ne of the reasons why the Chinese garden is such a potent symbol for the Chinese themselves must be the fact that it is one aspect of Chinese culture which has been widely admired in the West over several centuries, and which has actually had a significant influence on Western culture." This would certainly explain the initial popularity of the idea within Edmonton's Chinese community. She goes on to caution against an oversimplification of the idea of a Chinese garden; "there is," Hardie writes, "a tendency to essentialize the garden and regard it as a constant in Chinese culture, although it is very easy to demonstrate that in fact 'the garden' has meant many different things at different times or places, or in different circumstances."[17]

No one has come to appreciate this notion more than Francis Ng, who confronted several challenges to accepted tradition when he worked with landscape architect Patrick Butler to design the Edmonton garden.[18] The first of these was the sloping fan-shaped site offered by the city, not the traditional square or rectangular shape preferred by designers. Whereas water is generally accepted as an essential element of a traditional Chinese garden, geotechnical

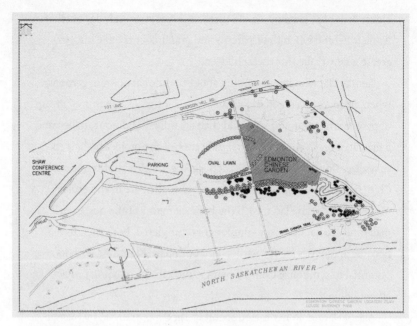

Edmonton Chinese Garden location plan, Louise McKinney Park, 2004.
[Drawing by Francis Ng and Patrick Butler]

conditions on the slope dictated that no water be used at the site. To accommodate the funding model, which consisted of ongoing efforts to raise money from a combination of public and private sources, the project had to be designed so that it could be executed in stages. Building materials chosen had to stand up to harsh climatic conditions and, to the degree possible, resist vandalism. And, perhaps most importantly, the location of the garden in a public park dictated a barrier-free, open, and unsupervised plan, a direct contradiction to a tradition of enclosure.

The resulting design, according to Ng, was defined by two axes, one running from southwest to northeast and the other from southeast to northwest. The architectural elements of the garden were located within the area defined by these axes, providing the visitor with spots to stop on his or her tour through the garden. The main focal point of the garden, and its chief architectural element, is a two-storey hexagonal pavilion (or *ting*) with a golden-yellow roof and curving, upturned

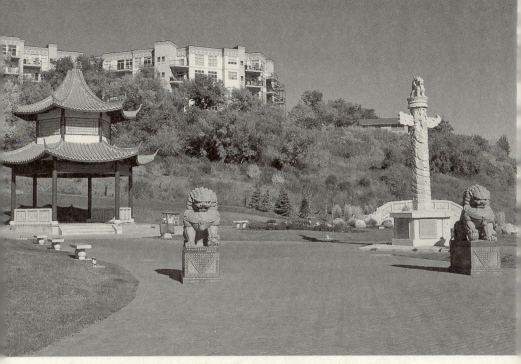

Edmonton Chinese Garden, 2007. [Photo courtesy of ISL Engineering and Land Services]

Artist's conception of the Edmonton Chinese Garden, 2004.

[Drawing by Tat Ming Yee for the Edmonton Chinese Garden Association]

eaves typical of those created during the Tang dynasty in China. There is a Chinese saying, says Ng, that "once a place has a ting we can call it a garden." Material for the pavilion and the expertise required to build it came from China. From it, visitors and those who go there to sit and contemplate can see the North Saskatchewan River "flowing towards the garden from the southwest." "This bright and colourful composition shines in the summer sky and adds a dash of colour in the winter snow," writes Ng.[19]

If the pavilion is the chief focal point of the garden, it is complemented by a stone bridge that arches over a rock pond, representing the friendship between the twin cities of Edmonton and Harbin, China. Surrounding the pond are the twelve zodiac figures from the Chinese calendar. A five-metre-tall stone pillar, adorned with a fiery dragon and surmounted by a dog-like "sky-howler," points to Beijing and recognizes the achievements of Chinese pioneers. At the base of the pillar is a dragon stone panel. The entrance to the garden is flanked by a giant pair of green rock lions, a gift from the People's Republic of China and a traditional sign of welcome for visitors. Yet to be funded and built are a main gate, a nine-dragon wall on the east side of the pavilion, and a moon gate at the southeast entrance to the garden.

Had they been given to indulgent fantasies, it is possible that Chin Lock, Jung Suey, Gee Gut, Hong Lee, Hop Woo, or Bark Ging Wong might have dreamed about such a thing as a traditional Chinese garden standing in the North Saskatchewan River valley, but they would have kept such fantasies to themselves. How might they have reacted to its being there? In their old age, Bark Ging and Young See Wong might have come occasionally to sit quietly under the upturned eaves of the pavilion while they watched their grandchildren compete in the annual dragon boat races taking place on the river below. And, if they had closed their eyes, they might have seen a double vision, one of a "veritable Chinese tapestry worked out in rectangles of harmonious greens, sea green, sage green, olive green, and the delicate apple green of lettuce beds" that reminded them of what they had accomplished in a foreign land and another of pavilions with upturned eaves, stone bridges, and dragon walls that reminded them of the rich heritage they had contributed to their children's homeland.

Clean up

9 Citizen Gardeners

Caring is the soul of gardening.
—LOIS HOLE, *I'll Never Marry a Farmer*

ONE MORNING IN MAY 1925, they could be seen scampering across the pages of the *Edmonton Bulletin* bearing hoes, rakes, ladders, and paint cans—stick figures against a silhouetted city skyline. A week earlier, they had been noticed in the *Edmonton Journal*, standing to attention just behind their officers, garden implements at the ready. The officers, scrub brush and paintbrush, were more menacingly armed with rifles, pistols, and swords. It was in this cartoonish fashion that Edmonton's two daily newspapers chose to advertise the 1925 annual citywide cleanup campaign. Implicit in the graphics was the message to citizens that by working together, each individual doing what he or she could to improve the appearance of the city, the goal of making Edmonton a clean and attractive city in which all could take pride was achievable. Organizers of the campaign, and of others like it, wanted their fellow citizens to care enough about the city to do their

Illustration for Edmonton's annual cleanup campaign, EJ, Saturday, May 8, 1925.

part to make it a better place. Gardening may be only one expression of this caring mentality, but, for the citizen gardener, it is important not only for its effects on the physical aspect of the city but because, as the ultimate form of nurturing, it carries over into all aspects of life.[1]

The organizers of that 1925 annual Clean Up, Paint Up, Plant Up campaign included Gladys Reeves, who acted as its official secretary. Reeves, one of Edmonton's first citizen gardeners and one of the most prominent in the city's gardening history, makes a good study as the prototypical citizen gardener. First, although she loved gardens and knew as much, it would seem, as her contemporaries about them, her importance to Edmonton's gardening history lies not at all in her gardening expertise or in the gardens she created, for she created none that are known about. Rather, her importance lies in her work as a gardening advocate. Promoter of city beautification through gardening and tree planting; volunteer organizer with groups such as the Edmonton Horticultural Society, the Edmonton Tree Planting Committee, the Alberta Horticultural and Forestry Association, and the city's annual cleanup campaign; writer of many

letters to newspaper editors and city officials about some aspect of city beautification; speech maker on subjects related to gardening and tree planting, Reeves thought of gardening as the ultimate form of civic engagement.[2] Gardeners were, to use a word she applied to her father, "upbuilders" of the city; her job was to make others understand, as she did, their importance.[3]

Reeves was firmly opposed to passivity in civic affairs; "if we want things done," she argued, "we must put our shoulder to the wheel and help to do them." In undated notes for one of her many speeches, she claimed that all levels of government were influenced in their policies by public opinion, the implication being that citizens must take every opportunity to express their opinions publicly. It was the "continued efforts" of the horticultural society, the tree-planting committee, and Edmonton's community leagues, she believed, that had influenced the city to increase its expenditures on parks and boulevards. Together, they had "awakened" the "city fathers" to the notion "that packing plants and business houses are not the only requirements of a city and are not the things which make and hold good citizens." And they provoked this change of policy not simply by voicing their collective opinion but by joining forces to work, as volunteers, with the city engineering department and the land department on programs of city beautification.[4]

For Reeves, and for all citizen gardeners, gardening and tree planting were effective agents for social change, especially if they were passed on as skills to children. She referred to schoolchildren as "citizens of tomorrow," open to absorbing lessons that would last them a lifetime. Although she had no children herself, she was a fond aunt to her nieces and nephews and encouraged them to enter horticultural competitions, she made photographing children a specialty of her business, and she often spoke to children in schools about tree planting. She was frequently called upon to judge school ground beautification competitions. In 1924, while she was president of the horticultural society, the profile given to children's classes in the annual show was increased substantially, with one afternoon devoted entirely to special children's activities and classes. The highlight of children's day that year, and for years afterwards, was a parade of tricycles, scooters, doll carriages, hobby horses, kiddie cars, bicycles, and children's "autos," all "prettily decorated," with prizes awarded in several categories and age groups. Eight-year-old Charles Dixon won a third prize in 1924 for his "clever float" entitled "The vacant lot pays"; loaded with produce from the garden, it featured two giant sunflower headlamps.[5] In the 1926 show, Reeves's nephew Hugh Reeves received a second prize for his wagon, on which he had constructed a miniature house with a garden round it and a sign saying, "every garden means a home."[6] For all that Reeves earned a reputation over her lifetime as both a critic and activist, her nurturing side came to the fore with children who, as she was known to say, "will only have the inspiration which we pass on to them."[7]

To be effective, to draw others into their vision, citizen gardeners must have qualities of leadership. They must be good communicators who project, in some measure, a personal charisma. Their energy, resourcefulness, and imagination make them good problem solvers and occasional projectors. But only some of them, and Reeves was one of these, develop a public presence that brings attention to causes. While tree planting may have been the most important of the many causes she took on, she managed to garner attention for almost any project devised in the name of city beautification. In October of 1931, for instance, a columnist in an Edmonton newspaper wrote about an "interesting idea" put forward by the horticultural society to create a floral colour scheme for the city, one that could be incorporated into public and private gardens, on roadway verges, and even on boulevards. Chosen by the society to put the "blue and gold" project into effect was Reeves, described by the columnist as a most suitable choice and as a "practical force" on the tree-planting committee. Reeves worked with officials from the City of Edmonton and with community leagues. Early in 1932, the Cromdale Community League planted blue and gold flowers on the embankment of the new bridge over the Rat Creek ravine. In August 1932, the *Bulletin* ran a story under Reeves's photograph headlined, "Plan to make city of blue and gold." According to the story, it was Reeves who had come up with the novel idea of creating this particular colour scheme for the city's public plantings. The blue was to come from delphinium, a perennial with attractive blue flower spikes. Potential disadvantages of delphinium, such as its height, vulnerability to wind, and toxicity to animals and humans, were omitted from mention in the article. The gold was to come from poppies, perhaps a better choice for boulevards because they reseed themselves and are possibly more resistant to trampling. To obtain seed, she wrote to the manager of Chateau Lake Louise, asking for seeds from the many yellow poppies grown on the hotel grounds, a request he responded to by sending "a few packages" and an explanation for why he could not send more. Had the blue and gold idea been put forward by anyone else, it is doubtful that it would have made the modest splash Reeves was able to achieve for it. Not surprisingly, the plan had limited success. But the fact that it was attempted at all was characteristic of Reeves, who devised and implemented a score of novel schemes intended to promote city beautification, most of which achieved the goals set for them.[8]

Citizen gardeners tend not to be loners; they work best and achieve most as key members of a team. Reeves may be the best-known citizen gardener of her times but all around her, and supporting her efforts, was a group of like-minded individuals. George Harcourt, professor of horticulture at the University of Alberta, saw in gardening a way to build not just a more attractive city but a better one. In his position with the Faculty of Agriculture, Harcourt rendered what assistance he could to the tree-planting committee. He also worked with the horticultural society on the vacant lot gardening program and promoted home gardening in radio broadcasts and publications.[9] George Buchanan, an employee of Edmonton City Dairy from 1916 until his death in 1939, worked with Reeves on the tree-planting committee and became known and respected for his work on the annual cleanup campaign. In May 1930, for instance, his photo appeared on the front page of the *Bulletin*, where he was described as sure to be "one of the busiest men in Edmonton in the immediate future having been named chairman of the City and Home Beautification Campaign committee."[10] William Cardy, who succeeded Reeves as president of the horticultural society in 1925 and who worked indefatigably for the next thirty-five years to stimulate an interest in gardening and to bring information and expertise to the gardening public, was possibly the least known citizen gardener of his own time but one of the most effective in the long term. Superficial differences aside, differences of age, gender, and profession, Reeves and many of those she associated with shared a belief in the power of horticulture to transform both a city and its citizens and a commitment to do what they could to make Edmonton a better place.

The transformative power of horticulture presents itself as a theme in Edmonton's gardening history, a theme carried forward from generation to generation by a succession of citizen gardeners, men and women for whom the garden is not a private refuge from the world but a model for action in it. Lois Hole, who settled on a farm in

St. Albert in 1952, twenty-two years before Reeves's death, was one of these. But while Reeves had been active during Edmonton's formative years, when city beautification was both a practical necessity and a convenient ideology, Hole's Edmonton was a rapidly growing metropolis adjacent to her home community of St. Albert. After eight years of struggling with her husband, Ted, to run a mixed farm in St. Albert, horticulture proved to be quite literally and materially transformative when the young couple decided to adapt their business model to an increasingly urban market by switching from grain and animal production to horticultural crops, first field vegetables sold at the farm gate and to local businesses, and then plants and other supplies for the home gardener. Hole, who loved gardening almost as much as she loved people, delivered a quality of service and product sought out by gardeners from Edmonton and its surrounding communities, including, of course, her home community of St. Albert. It was as entrepreneur par excellence that she began her own transformation from farm wife to beloved Lieutenant-Governor of Alberta, a position she held from February 2000 until her death in January 2005. Profoundly influenced by her love of plants and of gardening, she plucked from her own garden the maxim that informed her approach to public service, "caring is the soul of gardening."[11]

Hole's reputation as a gardener was hard earned and never forgotten. It was an indelible part of her persona. The gradual process of acquiring it began during her girlhood in small-town Saskatchewan in her mother's garden, where she loved to go with a best friend to pick a few tomatoes for eating out of hand. They took with them a cup of sugar in which to dip the freshly picked tomatoes before popping them into their mouths. It was a childhood memory, she wrote, that she would never forget, and its relevance to the table was not unimportant.[12] For Hole, gardening and the pleasures of the table were intimately related. Perhaps this was why, when she and her husband, Ted, began farming, what began as accidental and occasional sales of vegetables from her home garden seemed a natural fit. Gradually, the business evolved and changed its focus, moving from mixed farm to vegetable farm until it had grown to be one of the largest greenhouse and garden businesses in western Canada, and always Hole was at the centre of it. She became its brand and, by the 1980s and 1990s, the brand seemed almost to be a part of Edmonton's identity. Hole did not have to promote gardening in Edmonton and its surrounding communities, because, as combined entrepreneur and garden writer, whose books on gardening were an immediate popular success, she had come to represent it. And, in a fashion that was characteristic of her and endeared her to the public, she appeared to take the garden and the values she had acquired there with her into a parallel life of public service.

Hole's values are perhaps best conveyed in her autobiographical book, *I'll Never Marry a Farmer*, billed on the front cover as a book on life, learning, and vegetable gardening. In it, she writes primarily about the life she and her husband, Ted, made together on a beautiful piece of property overlooking the Sturgeon River in St. Albert. The picture she paints is a happy one, one in which the sense of triumph associated with overcoming adversity always stood out in her memory, relegating to the background the times of frustration and financial stress that resulted from farm disasters of one sort or another. Benchmark dates in the evolution of the business, such as 1960, when the Holes decided to abandon their idea of a traditional farm in favour of a vegetable farm and market garden business, or 1979, when the Holes' sons, Bill and Jim, joined the family business and set about to expand it, are downplayed in favour of different kinds of benchmarks, like the "light bulb" realization that the vegetable garden was not just a sideline or an amenity but the path forward, or the realization that a former staff person, a Métis woman named Virginie Durocher who had worked for the Holes intermittently, had, in her own quiet way, taught Hole as much about life as she had about gardening.[13] *I'll Never Marry a Farmer* stands as a personal record of Hole's married life, an informal history of the creation of a business, and an implied philosophy framed around the view that gardening is more than an activity; it is a metaphor for life.

As is the case with life, the chronological framework in *I'll Never Marry a Farmer* is richly decorated with events, memories, humorous incidents, and, in this case, by what was growing in Hole's garden. Topics succeed one another in a seemingly random order, each one named, each one illustrated, and each one clothing a lesson in the fabric of Hole's life. They are little fables. If soil is the medium for healthy plants, and if some plants thrive in one type of soil while others thrive in another type, so people need to know the kinds of situations in which they are most likely to flourish. Chapter 6, "Care and Nurture," proposes a parallel between a garden and a life. Both are loaded with risks that we take account of and then do what we can to minimize. Success is never assured; "[y]et, in the end," she writes, "we are almost always rewarded with a beautiful harvest."[14] Cucumbers, in another piece, are likened to babies; "they need constant love and attention to grow up happy."[15] An anecdote about a neighbour who "spread the gospel of trees" begins doubtfully but concludes with an endorsement of the doctrine of tree planting and an illustration of its benefits. A short disquisition on the "Golden Rule," do unto others as you would have them do unto you, is illustrated with a story about the unnecessary humiliation of a staff person by a customer over the weight of a bag of cucumbers. The story of the death of a staff member's son, a young man who had occasionally worked at Hole's but whose many problems prevented him from realizing the potential Hole recognized in him, is more than a personal tragedy; it is "Society's Loss." Distressed by this death, Hole had no solutions to offer; characteristically, however, she knew that "we can't afford to simply throw our hands in the air and lament that there's nothing we can do."[16] And, finally, she writes about the last vegetable harvest in 1991 as a sad, if inevitable, day in her life. Hole valued the "thousand priceless lessons" she had learned while working the land, but the most important lesson it taught her, she claimed, "is simply that there's still so much more to learn," a lesson she refers to as "the true harvest."[17]

Learning, for Hole, was a never-ending process. As for knowledge, there were two ways to acquire it, by doing or by studying. She lamented a tendency she had noticed in some people to favour one approach over the other; "[i]f there's anything I've discovered over the years, it's that both kinds of education are necessary for success."[18] Hole's devotion to the cause of education—one that began as early as 1967, when, unhappy because French was not being taught at her sons' school, she ran for, and won, a seat as a trustee for the Sturgeon School Division—remained with her throughout her life. More than thirty years as a trustee or chairperson of either the Sturgeon School Division or St. Albert School Division, and eleven years on the Athabasca University Governing Council, culminated in 1998 with her appointment as chancellor of the University of Alberta. Although her tenure there was less than two years, she made a profound impression on those in the university community with whom she worked, particularly those in the Faculty of Arts who found in her a champion of the value of a liberal arts education. For Hole, educating the imagination of young people, helping them to think creatively and to understand themselves as part of a social and political system, was a noble calling. As a lover of music and of reading, as a great encourager of youth, and as a promoter of public libraries, she revelled in her role as titular head of the University of Alberta, and was delighted that the institution awarded her an honorary doctor of laws degree in 2000, the year she relinquished her role as chancellor to accept an appointment as Lieutenant-Governor of Alberta.

As Lieutenant-Governor, Hole gained a platform from which she could speak to all Albertans, and her speeches were occasionally a surprise to her audiences. Addressing groups on subjects of importance to them, the opening of a new facility, perhaps, or the celebration of an event, she generally found ways to incorporate subjects of importance to her. The need for tolerance and understanding, the importance of initiative, the value of diversity, maintaining respect for others, the value of an education: these were just some of the topics she wove into her speeches, leaving her audiences provoked, stimulated, sometimes bemused, and always admiring. Even when Hole's views differed from those of the Government of Alberta, she was able to deliver herself in a manner that challenged without offending. Hole's approach to her job as Lieutenant-Governor was not too different from her approach to her job as a gardener. She nurtured and cared for the people of Alberta as she nurtured and cared for the plants in her garden. And, with faith in the future, she looked forward to a glorious harvest.

In his book *The World We Want*, Canadian philosopher Mark Kingwell describes citizenship as "one of the profound categories that make us who we are, one of the crucial ways humans go about creating a life for themselves. Without it," he claimed, "we are cut adrift from each other—and from ourselves."[19] Hole's authenticity as a person and as a public figure derived from the fact that her gardening persona and her public persona were one and the same. If caring was the soul of gardening, it was also the soul of citizenship; Hole did not so much draw the comparison as live it.

As principal of horticulture for the City of Edmonton from January 1, 1995, to January 19, 2013, finding ways to bring "the garden" into public consciousness within the context of citizenship was more than a challenge for John Helder, it was a job description. Just over eighteen years in the position confirmed Helder in a belief that had been forming when he accepted the position, that horticulture is not what we see but how community members and groups interact with nature to enhance the urban environment. Nature, Helder believes, is a part of urban life and it needs to be part of the awareness of urbanites. And, although his understanding of the relationship between nature and the garden evolved somewhat over the course of a long career, his commitment to making them an integral part of urban life remained firm.[20]

No alternatives presented themselves during adolescence to challenge Helder's early inclinations towards a career in horticulture. Born in Holland at the end of the Second World War, one of his earliest memories is of the delight and excitement he experienced at the age of five when an uncle, who owned and operated a greenhouse, presented him with a pot of fuschia. After moving with his family to Canada in 1954, and settling in the Ontario village of Chatsworth, two elderly sisters who were neighbours encouraged Helder to help them in the garden and he soon came to associate the pleasure of the work with their openness and hospitality. As an adolescent, odd jobs weeding gardens and mowing lawns led to summer jobs in one or another of the many nurseries, greenhouses, and market gardens near his home. And, in 1967, after a year with a nursery where he learned and practised techniques of plant propagation, Helder enrolled in the Niagara Parks Commission School of Horticulture, an intensive residential program now known as the Niagara Parks Botanical Gardens and School of Horticulture.

The blend of theory and practice that has always distinguished the horticultural diploma program at Niagara Parks suited Helder's natural inclinations and left him with a pragmatic and incremental approach to problem solving. He received his Niagara Parks diploma in 1970 and worked for seven years, five of them as superintendent of greenhouses at the Niagara Parks Commission, where he managed the greenhouses, organized displays of exotic plant material, and taught and supervised students at the school. It was the range of his duties at Niagara Parks, Helder believes, that made him an attractive candidate for the position of curator at Edmonton's then newly opened Muttart Conservatory. Interviewed in January 1977, appointed in February, John relocated to the city by March to begin what turned out to be a productive career with the City of Edmonton. Approximately eighteen years at the conservatory, where his title changed from curator, to supervisor, and finally to director of the facility, left him firmly convinced that the garden resided less in the ever-changing spectacle of the show pavilion than in the hearts, minds, and hands of those who came to use the facilities at the Muttart Conservatory and to participate in its programming. This was the philosophy Helder brought in 1995 to a new and undefined position as the city's first principal of horticulture.

Helder's approach to the city beautification file with which, as principal of horticulture, he was charged was never simplistic or prescriptive. His intent was always to engage individuals and groups to partner with the city on a wide variety of horticultural and environmental initiatives and to move, step by step, towards a particular goal; when there was disagreement among stakeholders over conflicting conceptions of beauty, or around the most appropriate set of uses for a piece of property, he worked towards finding solutions in compromise. Providing support to community groups was sometimes difficult from within a large civic bureaucracy with its finite resources, frequent changes of priorities, and ongoing administrative reorganizations. Nevertheless, as he prepared for a January 2013 retirement, Helder could look back on several accomplishments that, taken together, altered the face of the city and opened the way for increased citizen participation in the ongoing project of city beautification.

Although Helder cannot claim to have originated Edmonton's Partners-in-Parks program, which was conceived and put into operation some time in the 1980s, his endorsement of the philosophy behind it had far-reaching consequences. The program expanded dramatically under his watch and the notion of a contract between city and citizen upon which it was based was later adapted by Helder to facilitate a major community gardening initiative in the city. In its Partners-in-Parks formulation, the contract is between the partner (individual or group) and the city, with the partner taking responsibility for maintaining a portion of public land, generally a section of boulevard or perhaps the flower beds in a public park. The partner agrees to follow environmental and other guidelines and receives certain benefits in return, benefits conferred through the Edmonton Horticultural Society, which in turn partners with the city to further its own mission, that of promoting an interest in horticulture among Edmonton's citizenry. The city's many Partners-in-Parks, and they numbered approximately 250 in 2012, thus become agents of city beautification, creating eye-catching beauty spots that would be beyond the resources of the city to deliver itself.

Helder's interest in community gardens, piqued in 1997 when he was called upon by a group of community garden enthusiasts to help a Mill Woods co-op housing project secure access to a piece of public property, has been both professional and personal.[21] In his professional capacity, he immediately recognized the movement's potential to engage a large number of people in communities across the city in horticultural projects of their own making. If, as principal of horticulture, his job was to encourage citizens to make nature and the garden integral to their concepts of urban living, community gardening was one obvious way to do it. Thus, Helder did what he could to facilitate the movement, first by adapting the Partners-in-Parks agreement to a community garden context, and then by assisting the organizers of the movement to obtain start-up funding. In 2012, as a program operating under the umbrella of Sustainable Food Edmonton, approximately eighty community gardens were located in communities across the city. Intensively cultivated, often somewhat chaotic and overgrown in appearance, they have occasionally provoked complaints from neighbouring property owners who take issue with some aspect of their appearance. At the same time, they have become recognizable and defining features in their home communities and valuable gathering places for urban gardeners. For Helder, who has donated personal volunteer time to ensuring that the community garden movement in the city succeeds, a community garden is a pure expression of horticulture as civic engagement.

One of Helder's first major initiatives after becoming Edmonton's principal of horticulture was to involve the city in the annual Communities in Bloom events. Participation in Communities in Bloom, he must have reasoned, would be a way to raise consciousness both inside and outside the horticultural community about the many roles played by nature and the garden in urban life. In 1997, after attending the Communities in Bloom award ceremonies in St. John's, Newfoundland, and learning of Kitchener, Ontario's front yard recognition efforts, he returned with an idea for a complementary event, one that has taken shape in Edmonton as the annual Front Yards in Bloom awards, organized by the Edmonton Horticultural Society with the assistance of the city and Canada Post. These two events, separate but interrelated, may have done as much to promote gardening in the city and to support a variety of horticultural and environmental initiatives as any other single initiative. Packaged for twenty-first-century sensibilities with awards ceremonies and declared winners, they are the contemporary equivalent of the early-twentieth-century efforts to recruit citizen armies that would voluntarily march to the tune of clean up, paint up, fix up, and plant up.²²

For Helder, "nature" and "the garden" are not opposing concepts, with nature representing a balanced, biodiverse, and self-sustaining ecosystem, while the garden is conceived as a static and artificial construct requiring human intervention to survive; for him, nature comprehends the garden. Hence, it is not surprising that the Naturescapes program, another of the programs that Helder helped to put in place during the early years of his mandate as principal of horticulture, uses the word "naturescape" to replace the word "garden." By partnering with Edmonton Catholic Schools, Edmonton Public Schools, and subsequently the francophone Conseil scolaire Centre-Nord, the City of Edmonton Department of Community Services, with the principal of horticulture as a resource person, provides support to school projects, each one of which is designed by the children and their teachers to serve a particular set of requirements. Depending on the age and interests of the children, a naturescape can be a butterfly garden; a bird habitat; a tree and shrub garden, otherwise known as a nursery; or perhaps a garden to supply vegetables for a school lunch program. Alternatively, it can be a project that attempts to create a particular landscape such as a wetland, a woodland, or a prairie grassland. A naturescape project can be built around the concept of stewardship, where children undertake to examine a habitat near their school with a view to identifying and removing invasive species. All naturescape projects involve the care of a defined piece of ground, but each one is individualized to serve a specific set of community and curricular goals. As society's interests have shifted towards the importance of ecology and the need to create sustainable communities, naturescape projects have become useful outdoor classrooms for teaching these concepts to children.

Helder's personal inclination towards living with nature rather than trying to control it is reflected in, and echoed by, many of changes that have occurred within the city since he took office. A Master Naturalist Program, whereby citizens are trained to become volunteer stewards of

Illustration for Edmonton's annual cleanup campaign, EB, Saturday, May 16, 1925.

nature in a variety of community settings, was conceived by Helder and is delivered by the Office of Biodiversity. Naturalization, a vegetation management strategy that privileges native ground covers and trees over a combination of sod and non-native species planting, has resulted in the reappearance of native spruce and aspen trees along highway and freeway verges. The city has reduced its use of herbicides, leading to the reappearance of dandelions and other so-called weeds on boulevards and in parks. For citizens accustomed to formal landscaping and manicured boulevards and parks, these changes can be disconcerting. But for the Edmonton Native Plant Group (formerly the Edmonton Naturalization Group), another of the many partner groups with which Helder has cultivated a working relationship, most of these changes have been welcome.

Helder's relationship with the Edmonton Native Plant Group, which began early in his tenure as principal of horticulture, resulted in many successful and ongoing partnership agreements. The group, an informal collection of individuals committed to the stewardship of native plants and to promoting their use in gardens, undertook to carry out a variety of projects, some aimed at restoring defined geographical areas to their natural state by removing invasive species and others designed to educate. The first of the group's restoration projects to get underway as a Partners-in-Parks project was the Mill Creek Thistle Patrol. Efforts by

individuals in 1997 and 1998 to replace the use of pesticides and herbicides in parts of Mill Creek Ravine with hand labour to remove noxious weeds led them to Helder and the eventual adaptation of the Partners-in-Parks agreement to their unique project. The Mill Creek patrol, which continues to operate, became a model for similar projects in other parts of the city. The native plant garden at the John Janzen Nature Centre, on the other hand, established in 2003 as a Partners-in-Parks project and maintained from year to year on the same basis, was designed to educate visitors to the centre regarding native plants suitable for home gardens. And, finally, through a Partners-in-Parks agreement, the group assumed responsibility for a portion of the City Tree Nursery, growing close to two hundred native plants either for their seed or to supply city naturalization projects and school naturescape projects. Not since the days of Gladys Reeves and the tree-planting committee had a group of citizens been granted planting privileges at the city nursery in exchange for a supply of planting material for public areas.

Edmonton has had, and will continue to have, a tradition of citizen gardeners, individuals whose participation in the life of the city and whose major contributions to the quality of urban life have been informed by the garden or channelled through it. Indeed, as a result of Helder's creative and participatory approach to his position as principal of horticulture, citizen gardeners in Edmonton have proliferated, empowered by policies and procedures set up to facilitate their participation in the urban project.

Kingwell ended his book *The World We Want* by saying that "[t]he times could not be riper for a playful but serious utopianism, an open-minded but political consciousness, an imaginative but hard-headed idea of civic participation." "How else," he wondered, "can we hope to fashion the world we want?"[23] It is a question that conjures up an image, an image that resembles the one published in the *Edmonton Bulletin* in May 1925. Animated stick figures with hoes and trowels populate the foreground. Behind them is the city and somewhere within it and around it is the garden, never quite perfect, except in the imagination of the citizen gardener.

Notes

Preface

1. The Edmonton Horticultural Society, in existence in Edmonton since its formation in 1909, changed its name in 1918 after amalgamating with the Vacant Lots Garden Society. From 1918 until 1973, when the original name was reinstated, the society went by the official name, "Edmonton Horticultural and Vacant Lots Garden Association." Nevertheless, it was often referred to simply as the Edmonton Horticultural Society even by its members and officers and, in this book, I have stuck to the original name, Edmonton Horticultural Society.
2. The website of the University of British Columbia, Indigenous Foundations can be found at http://indigenousfoundations.arts.ubc.ca.

1 Why Grow Here?

1. "Local," quoting the remarks of a Mr. Walpole Rolland, C.E., topographical draftsman for the Canadian Pacific Railway Company, *EB*, July 7, 1883, 1.
2. "Edmonton," *EB Supplement*, March 6, 1892, 1.
3. Editorial, *EB*, November 5, 1881, 1.
4. "Edmonton agricultural exhibition," *Saskatchewan Herald*, November 17, 1879, 2.
5. "The show," *EB*, October 28, 1882, 2.
6. "The exhibition," *EB*, October 9, 1886, 2.

7. "More samples," *EB*, September 26, 1891, 2.
8. "Local," *EB*, August 25, 1888, 1.
9. Thomas Henderson arrived in the Edmonton area in 1881, settling first on a farm at Little Mountain (the present-day area of Brintnell in northeast Edmonton), moving in 1883 to a home on Fraser Avenue (98th Street) in the city, and then again in 1891 to a farm at Rabbit Hill (now part of southwest Edmonton). His success with beekeeping, which he took up seriously while living on Fraser Avenue and continued after he moved to Rabbit Hill, was followed carefully in the *Edmonton Bulletin*. Fort Edmonton Park contains replicas of the Rabbit Hill home and the original round barn built by Henderson at Rabbit Hill in 1898.
10. "Fruit trees," *EB*, September 22, 1883, 2.
11. "Fruit," *EB*, September 8, 1888, 3.
12. "Local," *EB*, May 10, 1890, 1.
13. "Local," *EB*, July 6, 1889, 1.
14. "Local," *EB*, June 27, 1891, 1.
15. "Our vegetable garden," *EB*, August 14, 1893, 4.
16. Ross opened the Edmonton Hotel in 1876.
17. Gerald Friesen, *The Canadian Prairies: A History* (Toronto: University of Toronto Press, 1984), 249–50.
18. Gladys Reeves, notes for a speech to an unidentified group, n.d., PR1974.173/39b, PAA.
19. Gladys Reeves, letter to Mr. Wallace, Managing Editor of the *Edmonton Journal*, n.d. [October 1933], thanking him for "space given in your paper to a write-up of my old Dad W.P. Reeves," PR1974.173/33, PAA. The article Reeves referred to was "Active at 89, citizen says moderation longevity key," *EJ*, October 26, 1933, 15. A photo of W.P. Reeves taken by Gladys accompanied the article.
20. W.P. Reeves, letter to Messrs. Carters Tested Seeds Ltd., Raynes Park, London, March 27, 1935, PR1974.173/237, PAA.
21. Gladys Reeves, letter to Mr. Wallace, Managing Editor of the *Edmonton Journal*, n.d., PR1974.173/33, PAA.
22. "Fruit trees," *EB*, September 22, 1883, 2.
23. "Garden expert, W.P. Reeves, to celebrate 95th birthday," *EJ*, October 20, 1939, 22.
24. "Fruit trees," *EB*, September 22, 1883, 2.
25. Editorial, *EB*, March 21, 1885, 2.

26. Georges Bugnet, *The Forest*, trans. David Carpenter (Montreal: Harvest House, 1976), 117.
27. Jean Papen, *Georges Bugnet: homme de lettres canadien* (Saint-Boniface, MB: Éditions des Plaines, 1985), 37. The French reads: "à une âme chrétienne comme la sienne, l'horticulture lui permet, grâce au contact intime et constant avec les lois secrètes et grandioses de la nature, d'admirer la sagesse divine qui s'y dévoilait." ("To a born Christian like Bugnet, horticulture, because of its intimate and constant contact with the secrets and all-powerful laws of nature, provided a glimpse into the unfolding of divine wisdom") (author's own translation).
28. The 'Thérèse Bugnet' rose was named after a younger sister who came to Canada with two other members of the family in 1908 but was escorted back to France in 1914 by Charles Bugnet, brother to Georges, where she entered the Carmelite Order. See Papen, *Georges Bugnet*, 34–35.
29. "Local," *EB*, August 23, 1890, 1.
30. "Strathcona flower show a big success," *EB*, August 10, 1911, 2.
31. These articles were written under his byline, "The Gardener," and appeared weekly on Saturdays throughout the growing season of 1932. See, for example, the columns in the *Bulletin* on April 30, 1932, 3, and May 7, 1932, 7, which focus on roses.
32. For a fuller account of the Capitol Theatre rose shows, see the seventh essay, "Edmonton, the Rose City." When Walter Wilson left the Capitol Theatre to manage the Paramount Theatre, the annual rose shows moved with him. Wilson retired in 1954, but in 1955 Hilda McAfee won a trophy for the best rose in the Paramount Theatre Rose Show. This cup, along with another cup and a small silver bowl, are in MS 226.2, File 103, CEA.
33. *Farmstead Planning and Beautification*, Publication No. 9 (Edmonton: Alberta Department of Agriculture).
34. George Shewchuk, interviews by author, July 4 and 24, 2002; August 9 and 27, 2002; October 24, 2002.
35. Jim Hole, *What Grows Here*, vol. 1, *Locations* (St. Albert, AB: Hole's Publishing, 2004).

2 Donald Ross: Edmonton's "Father of Gardening"

1. "Donald Ross' greenhouses," *EB*, Saturday, January 3, 1903, 1.
2. John Patrick Day, "Donald Ross, Raconteur and Entrepreneur," CEA library; John Patrick Day, "Donald Ross, Old Timer Extraordinaire," in *The Life of a City*,

ed. Bob Hesketh and Frances Swyripa (Edmonton: NeWest Press, 1995), 31–39; Archibald Oswald MacRae, *History of the Province of Alberta*, vol. 1 (Calgary: The Western Canada History Co., 1912), 541–43; "Settlers' experience," *EB*, April 26, 1890, 3.

3. Day, "Donald Ross, Old Timer Extraordinaire," 34.
4. MacRae, *History of the Province of Alberta*, vol. 1, 543.
5. Day, "Donald Ross, Old Timer Extraordinaire," 1; "Old Edmonton hotel burned: loss $25,000," *EB*, May 11, 1928, 1 and 15.
6. Advertisement for the Alberta Hotel, *EB*, May 6, 1882, 3.
7. "Local," *EB*, July 5, 1884, 1; "Local," *EB*, September 17, 1887, 1.
8. Information about gardens and field crops grown at Fort Edmonton comes from travellers' accounts and from journals maintained at the fort by Hudson's Bay employees, the latter of which are located in the Hudson's Bay Company Archives located in Winnipeg, Manitoba. See particularly the Journals, B.60/a/1–42, and District Reports, B.60/e/1–24. George M. Grant, *Ocean to Ocean, Sandford Fleming's Expedition Through Canada in 1872* (Toronto: James Campbell, 1873; facsimile reprint edition, 1970), 172, describes barley, potatoes, and turnips as "sure crops."
9. "Local," *EB*, April 29, 1882, 2; "Local," *EB*, May 6, 1882, 3.
10. "An agricultural society," *EB*, September 9, 1882, 2; "The show," *EB*, October 28, 1882, 2.
11. Day, "Donald Ross, Raconteur and Entrepreneur," 29. For references to Ross's vegetables, see the "Local" column in the *Edmonton Bulletin* on November 5, 1881, 1; September 23, 1882, 1; July 21, 1883, 1; August 4, 1883, 1; September 25, 1886, 4; October 2, 1886, 1; July 5, 1890, 5; May 23, 1891, 1; June 27, 1891, 1.
12. 'Beauty of Hebron' potatoes are described online as having rosy skin and white flesh. Peter Dreyer, *A Gardener Touched with Genius: The life of Luther Burbank* (New York: Coward, McCann & Geoghegan, 1975), 89.
13. "Local," *EB*, November 5, 1881, 1; "The exhibition," *EB*, October 21, 1882, 3; "The exhibition," *EB*, October 13, 1883, 2; "Local," *EB*, September 25, 1886, 4; "Local," *EB*, July 6, 1889, 1; "The exhibition," *EB*, October 19, 1889, 4; "Local," *EB*, June 27, 1891, 1; "Agricultural exhibition," *EB*, October 17, 1891, 2; "Our vegetable garden," *EB*, August 14, 1893, 4; "Local," *EB*, July 19, 1894, 1. The July 19, 1894, "Local" column in the *Bulletin* notes, "Luther Burbank of Santa Rosa, California, arrived on Monday's train. Mr. Burbank makes a business of originating new varieties of fruits and flowers by hybridization and other means, and is visiting Edmonton with a view to securing something new in these lines." The correct

spelling of 'Red Weatherfield' is 'Red Wethersfield', named for the town of Wethersfield, Connecticut. There are several varieties of both Dutch and express cabbage. 'Windsor' is the varietal name of a broad bean. There are many varieties of long pod beans.

14. "South Edmonton fall exhibition," *EB*, October 4, 1894, 2.
15. "Edmonton exhibition," *EB*, October 11, 1894, 4.
16. Advertisement for Renfrew Fruit and Floral Company, *EB*, March 7, 1885, 1; "Local," *EB*, April 11, 1885, 1.
17. "Local," *EB*, July 6, 1889, 1.
18. "Local," *EB*, July 5, 1890, 1.
19. "Local," *EB*, June 27, 1891, 1.
20. "Our vegetable garden," *EB*, August 14, 1893, 4.
21. "Local," *EB*, July 15, 1882, 1.
22. "Fruit trees," *EB*, September 22, 1883, 2.
23. "Fruit," *EB*, September 8, 1888, 3.
24. "Edmonton," *EB*, February 22, 1894, 3.
25. "Local," *EB*, June 28, 1894, 1.
26. "Mangolds," generally referred to by their whole name, mangold wurzel (or mangelwurzel), are root vegetables similar to a beet with leafy greens and are used primarily as cattle feed. The "citron" referred to in the newspaper may have referred to citron melons, a fruit grown primarily for its rind that was candied and used in baking. "An agricultural society," *EB*, September 9, 1882, 2; "The agricultural society," *EB*, September 30, 1882, 3; "The exhibition," *EB*, October 21, 1882, 2.
27. "Agricultural exhibition," *EB*, October 18, 1890, 1.
28. "Edmonton exhibition," *EB*, October 11, 1894, 4.
29. "Local," *EB*, February 18, 1895, 1.
30. "Edmonton exhibition," *EB*, October 7, 1895, 3.
31. "South Edmonton exhibition," *EB*, October 10, 1895, 3.
32. "For Calgary fair," *EB*, September 14, 1900, 2.
33. "Western horticultural fair," *EB*, September 12, 1902, 2.
34. "Donald Ross' greenhouses," *EB*, January 3, 1903, 1.
35. "Local," *EB*, August 15, 1891, 1.
36. "Local," *EB*, July 19, 1894, 1.
37. Advertisement in *EB*, July 10, 1893, 1.
38. "Local," *EB*, July 13, 1893, 1; "Local," *EB*, May 14, 1894, 1.

39. References to Frank Marriaggi (sometimes spelled Mariaggi) are numerous in the *Edmonton Bulletin* during the years he managed the Alberta Hotel (1893–1895). I have chosen the spelling with a double *r* and a double *g* for two reasons: first, because it is consistently used in the advertisements that Marriaggi must have composed; and second, because it gradually supersedes the other.
40. Advertisement for the Alberta Hotel, *EB*, August 31, 1893, 4.
41. "Local," *EB*, May 18, 1893, 1; "Local," *EB*, June 15, 1893, 1.
42. "Local," *EB*, August 6, 1894, 1.
43. "Local," *EB*, October 1, 1894, 1.
44. "Local," *EB*, July 15, 1895, 1.
45. "Local," *EB*, October 28, 1895, 1.
46. "Donald Ross' greenhouses," *EB*, January 3, 1903, 1.
47. "Donald Ross' greenhouses," *EB*, January 3, 1903, 1.
48. Day, "Donald Ross, Raconteur and Entrepreneur," 29–30.
49. "Local," *EB*, July 19, 1894, 1.
50. "Local," *EB*, July 25, 1895, 1.
51. Eight decades later, Edmonton lily breeder Fred Tarlton explained that *Lilium philadelphicum andinum*, "our sole native lily," had become somewhat scarce. He advised locals to pick it only when the target plant would otherwise be "doomed to certain destruction." Fred Tarlton, "The Planting and Transplanting of Lilies," *Kinnikinnick* 1, no. 3 (February 1975): 3–5.

3   Nature's Garden Transformed

1. Jane Brown, *The Pursuit of Paradise: A Social History of Gardens and Gardening* (London: HarperCollins, 2000), 317.
2. Clinton Evans, *The War on Weeds in the Prairie West: An Environmental History* (Calgary: University of Calgary Press, 2002), 190.
3. George H. Turner, "Plants of the Edmonton District of the Province of Alberta," *The Canadian Field-Naturalist* 63, no. 1 (January–February 1949): 1, http://www.biodiversitylibrary.org/item/89254#page/12/mode/1up.
4. Kathleen Wilkinson, *Trees and Shrubs of Alberta: A Habitat Field Guide* (Edmonton: Lone Pine Press, 1990); France Royer and Richard Dickinson, *Plants of Alberta* (Edmonton: Lone Pine Press, 2007).
5. Wilkinson, *Trees and Shrubs of Alberta*, 1–4.
6. J.A. McKeague and P.C. Stobbe, *History of Soil Survey in Canada 1914–1975*, Historical Series No. 11 (Ottawa: Canada Department of Agriculture, 1978), 7–8.

7. "Why," *EB*, July 8, 1882, 2.
8. "Why," 2.
9. Osler's account as it pertained to the Edmonton district was reprinted by the *Bulletin* in two installments: "Dundee Courier," *EB*, February 1, 1894, 3; and "Edmonton," *EB*, February 22, 1894, 3. "Mr. Osler's lectures on the Northwest in Scotland," the *Bulletin* reported in the local column of the January 29 edition, particularly recommended the Edmonton area to prospective settlers "with limited means" (1).
10. See, for example, Heinz W. Pyszczyk, Ross W. Wein, and Elizabeth Noble, "Aboriginal Land-Use of the Greater Edmonton Area," in *Coyotes Still Sing in My Valley: Conserving Biodiversity in a Northern City*, ed. Ross W. Wein (Edmonton: Spotted Cow Press, 2006), 21–47. On the basis of archaeological evidence, they conclude that "there was an Aboriginal presence here for many millennia." During what they refer to as the "Late Prehistoric Period" (about 1,750–300 years ago), "Aboriginal peoples (likely ancestral to Blackfoot) occupied the Greater Edmonton area." They identify Edmonton as "a key rendezvous point and staging area," and they use historic sources to support the theory that Cree had moved into the area "long before the permanent establishment of fur trade posts."
11. Anne Anderson, *The First Métis—A New Nation* (Edmonton: UVISCO Press, 1985), 32. David G. Mandelbaum, *The Plains Cree: An Ethnographic, Historical and Comparative Study*, Canadian Plains Studies 9 (Regina: Canadian Plains Research Center, University of Regina, 1979), 74–78 notes five basic mainstays of the Plains Cree economy: the buffalo, the horse, the dog, small animals, and vegetal foods. He names the various plants used by the Plains Cree and describes exactly how they were prepared.
12. W.R. Leslie, "Horticulture Among the Indians of the Northern Great Plains," chap. 2 in *Development of Horticulture on the Northern Great Plains*, ed. W.H. Alderman (St. Paul, MN: The Great Plains Region American Society for Horticultural Science, 1962), 6–7.
13. Patrick Seymour, interviews by author, July 24, August 11, and December 3, 2003.
14. Anne Anderson, *Some Native Herbal Remedies*, as told to Anne Anderson by Luke Chalifoux, Friends of the Botanic Garden of the University of Alberta, Publication Number 8 (Edmonton: Department of Botany, University of Alberta, 1977), 2.

15. Harold A. Innis, *The Fur Trade in Canada: An Introduction to Canadian Economic History*, rev. ed. (Toronto: University of Toronto Press, 1970), 299–301; Ted Binnema and Gerhard J. Ens, eds., *The Hudson's Bay Company Edmonton House Journals, Correspondence, and Reports: 1806–1821* (Calgary: Historical Society of Alberta, Alberta Records Publication Board, 2012) trace the expansion of garden and grain fields at the fur trading posts in the interior, including Edmonton House, to a HBC "reorganization and retrenchment policy" of 1810, "when directives were sent to all posts that the cost of European provisions, particularly flour, had to be reduced and that grain and vegetables should be grown at posts, whenever feasible" (72).

 The official name of the HBC fur trading post at Edmonton was Edmonton House. However, after about 1795, according to Binnema and Ens, posts along the North Saskatchewan River required fortification (30). This may account for the tendency to refer to Edmonton House as Fort Edmonton. I have used both terms in the text, depending on the context.

16. Binnema and Ens, *The Hudson's Bay Company Edmonton House Journals*, note that "[i]n exchange for the right to live at the posts and share in the food, these women and children worked for the companies. French, English (with Irish, Scots, and English accents), Cree, Assiniboine, Blackfoot, and Gros Ventre were all heard routinely in and about the settlement" (66).
17. "Edmonton Post Journal 1825–1826," B.60/a/23, HBCA.
18. "Edmonton Post Journal 1827–1828," April 18, 1828, B.60/a/27, HBCA.
19. "Edmonton Post Journal 1827–1828," September 28, 1827, B.60/a/25, HBCA.
20. "Edmonton Post Journal 1832–1833," B.60/a/27, HBCA.
21. G.1/92, 1846, HBCA.
22. B.6566, PAA.
23. Edmonton District Report, 1862, B.60/e/9, HBCA.
24. Grant, *Ocean to Ocean*, 172.
25. Georges Bugnet, *Nipsya*, trans. Constance Davies Woodrow (New York: Louis Carrier & Co., 1929).
26. Bugnet, *Nipsya*, 175–76. Bugnet is likely referring to bog cranberries (*Vaccinium vitis-idaea*) when he uses the word, more commonly used in French, *atoca*. Pembina is another name for highbush cranberry (*Viburnum opulus*). I believe *graines d'orignal* could refer to moose berries when translated into English.
27. Bugnet, *Nipsya*, 187–88.
28. Bugnet, *Nipsya*, 104.
29. "Local," EB, August 7, 1886, 3.

30. Advertisement, *EB*, May 6, 1882, 1.
31. "Edmonton district," *EB*, April 5, 1890, 2.
32. "Prof. Macoun's opinion," *EB*, November 23, 1883, 3.
33. Editorial, *EB*, March 21, 1885, 2.
34. "An experimental farm," *EB*, July 19, 1890, 2.
35. Canada Department of Agriculture, Canada Agriculture, *The First Hundred Years*, Historical Series No. 1 (Ottawa: Queen's Printer, 1967), 9–16.
36. "Local," *EB*, August 25, 1888, 1.
37. "Local," *EB*, July 6, 1889, 1; "Agricultural exhibition," *EB*, October 18, 1890, 1.
38. "Local," *EB*, September 5, 1891, 1.
39. "Fruit trees," *EB*, September 22, 1883, 2.
40. "Fruit," *EB*, September 8, 1888, 3.
41. "Local," *EB*, July 13, 1889, 4.
42. "Local," *EB*, August 29, 1891, 1.
43. "Glorious summer," *EB*, July 9, 1894, 2.
44. Frederick G. Todd, "Report [to the City of Edmonton] on Parks and Boulevards," April 1907, 3, Parks and Recreation History clippings file, CEA.
45. Todd, "Report on Parks and Boulevards," 6. In 1916, Rat Creek Ravine was renamed Kinnaird Ravine after George Johnstone Kinnaird, who was hired by the City of Edmonton in 1900 as the city's secretary-treasurer and one of its first three city commissioners.
46. Morell & Nichols, Landscape Architects, "A Report on City Planning," presented November 1912, Parks and Recreation History clippings file, CEA.
47. Morell & Nichols, "A Report on City Planning," 68, 71.
48. Paul von Aueberg, "Report of the Parks Department for 1912," in *The City of Edmonton Alberta Eighth Annual Financial and Departmental Report For Year Ended October 31st 1912*, 175–81, provides an excellent snapshot of von Aueberg's philosophy and of his work. See also, "City of Edmonton Parks Department, Looking Back from January 2, 1959," Parks and Recreation History clippings file, CEA.
49. Information about Edmonton's river valley parks system, including the *North Saskatchewan River Area Redevelopment Plan*, the *Ribbon of Green Concept Plan*, and the *Urban Parks Management Plan: 2006–2016*, can be found on the city's website, http://www.edmonton.ca.
50. Von Aueberg, "Report of the Parks Department for 1912," 177.
51. Von Aueberg, "Report of the Parks Department for 1912," 175–81. For information about the history of Borden Park, see Martina Gardiner, "Borden Park at

100," in *A Century of Gardening in Edmonton*, ed. Maggie Easton et al. (Edmonton: Edmonton Horticultural Society, 2009), 13–15.

52. *The City of Edmonton Alberta Eighth Annual Financial and Departmental Report For Year Ended October 31st 1913*, 248, CEA.

53. "Edmonton the beautiful," *Town Topics*, Industrial and Investor's Number, 1913, 37 and 44, MS 325, Class 3, CEA.

54. Gardiner, "Borden Park at 100," 14.

55. City of Edmonton, *Urban Parks Management Plan: 2006–2016*, 23.

56. Gladys Reeves, notes for a speech, n.d., PR1974, 173/39a, PAA.

57. On January 23, 1931, John F.D. Tanqueray, senior planner for the Town Planning Commission, wrote to the city commissioners recommending that a replacement bridge be built over Rat Creek at 82nd Street and 111th Avenue. Tanqueray explicitly recommended against the alternative of simply filling the ravine and building over it. In addition, he recommended expropriating six residential lots along the ravine to expand a city park there. This recommendation was put to the city council along with plans for a concrete bridge. A flurry of community response ensued, including a letter from the Cromdale Community League protesting the idea of a bridge and recommending instead that the ravine be filled. Letters for and against were plentiful, and it was not until December of that year that council finally decided to proceed with the bridge and authorized the city's solicitor to prepare the contract. Reeves's letter is not included in the commissioners' papers for that year, but a copy of it, probably sent to a newspaper, is in the Gladys Reeves fonds at the Provincial Archives of Alberta. See RG 11, Series 5, Box 2, File 23, CEA, and Reeves's letter to unnamed recipients, 1974.173/31, PAA.

58. Commissioners' Report No. 16 to the Aldermen, City of Edmonton, April 25, 1924, RG 11.1, Class 1, File 18, CEA.

59. Gladys Reeves, Secretary, The Edmonton Tree Planting Committee, letter to the City Commissioners, May 20, 1927, RG 11, Series 1, Sub-series 1.4, Box 20, File 472, CEA.

60. Gladys Reeves, notes for a speech, n.d., PR1974.173/39a, PAA.

61. George Harcourt, "Beautifying the Home Grounds," *The Press Bulletin*, issued by the Department of Extension of the University of Alberta, 15, no. 5 (May 9, 1930): 1–8. UAA, Accession No. 72-129:1–9. To note, the Department of Extension was later granted faculty status in the 1970s.

62. See note 49 in this chapter.

63. *Inventory of Environmentally Sensitive and Significant Natural Areas*, prepared by Geowest Environmental Consultants Ltd., can be found on the City of Edmonton's website at http://www.edmonton.ca/city_government/documents/Geowest_1993_Inventory_of_ESA_and_SNAs.pdf.
64. Joy Finlay and Cam Finlay, "Little Mountain on the prairie," *EJ*, August 19, 1994, D1.
65. On November 6, 1998, Ron Chalmers wrote a column for the *Edmonton Journal* headlined, "Conservation cash needed," in which he argued for the use of municipal funds to purchase and preserve natural areas such as Little Mountain. Allan Chambers followed the Little Mountain debate with articles in the *Edmonton Journal* such as "Proposed fund will be too late for Little Mountain, supporters say," February 6, 1999, B1; "Councillors urged to boost fund for preserve," February 9, 1999, B3; "Land swap proposed by city to save Little Mountain," February 23, 1999, B1; "City's $150,000 starts fund for parkland," March 3, 1999, B1; "Council votes to preserve nature," April 7, 1999, B3; "Little Mountain still has steep hill left to climb," August 31, 1999, B3; "Little Mountain wins delay in development," September 1, 1999, B2; "Mayor says cost determines sanctuary's fate," September 29, 1999, B7; "Gems in the rough," November 4, 1999, A1 and 20; "Plan for fund to save natural areas put on hold," November 23, 1999, B4.
66. Ron Chalmers, "Joyless decision on Little Mountain marred by mayor's remarks," *EJ*, November 3, 1999, B3.
67. *EJ*, November 8, 1999, A12. For an excellent account of the volunteer efforts to preserve Little Mountain, see Patsy Cotterill, "Losing Little Mountain: A Personal Anatomy of a Campaign to Save a Remnant Aspen Parkland Site," in *Coyotes Still Sing in My Valley*, ed. Ross W. Wein (Edmonton: Spotted Cow Press, 2006), 231–48.
68. Grant Pearsell and Angela Hobson, interview by author, January 17, 2010. Grant Pearsell, interview by author, January 13, 2014. Information about the Edmonton and Area Land Trust, accessed October 4, 2012, can be found at http://ealt.ca/.

    According to Pearsell, the City of Edmonton's financial commitment to the Edmonton and Area Land Trust, which included half a million dollars in start-up funding, a two and a half million dollar operational endowment, and backing for a twenty million dollar borrowing initiative on purchases, has been largely responsible for the land trust's effectiveness to date. In 2011, the city published its environmental strategic plan, *The Way We Green*, a document

Pearsell refers to frequently in his work. It can be accessed via the city's website, http://www.edmonton.ca.

69. Patsy Cotterill, telephone conversation with author, October 10, 2012. Work to restore pockets of nature to prominence in the gardens and public parks of the city is undertaken by various individuals and organizations, including the Edmonton Native Plant Group. Information about the group and its many projects can be found at http://www.edmontonnativeplantgroup.org.

4 Among the First

1. For full and reliable information about Ramsay's life and career, I have relied heavily on Jane McCracken, "Walter Ramsay Florist Ltd." (Edmonton: Edmonton, Parks and Recreation, Historical and Science Services, May 1976), Fort Edmonton Park Report 59, CEA. McCracken's report was researched and written prior to the recreation of the Ramsay greenhouse as part of 1905 Street at Fort Edmonton Park. In addition to primary sources such as newspaper articles, it contains material obtained through interviews with members of Ramsay's staff and family.

2. Much misinformation appears in published articles about Bugnet's life and horticultural achievements. The most thorough and reliable account, and the one I have relied on for this chapter, is Jean Papen, *Georges Bugnet: homme de lettres canadien* (Saint-Boniface, MB: Éditions des Plaines, 1985).

3. Information on the life and career of Pike is difficult to come by. A brief article about him and his career ("Gardening hobby made him pioneer") was published in an unidentified newspaper contained in a clippings file at the City of Edmonton Archives. Some information about him was conveyed to the author directly in two interviews with Pike's employee and then business partner Walter Dorin, November 26, 2003, and February 2, 2004. In 1975, Pike celebrated the sixtieth anniversary of his seed business and a short history of the business, presumably written by Pike himself, was included on the back page of that year's seed catalogue. In 1982, a few months after Pike's death, Pike & Co. Ltd. was sold to A.E. McKenzie Co. Ltd., a seed company operating out of Brandon, Manitoba. McKenzie Co. was kind enough to send me photocopied documents relating to the corporate and financial status of Pike & Co. Ltd. between 1953, when it restructured to include Alfred Pike Junior as his father's partner, and 1982, when the business was sold. Information about Pike's work with the Capitol Theatre rose shows comes from newspaper articles, while documentation for his work with the Edmonton Horticultural

4. Society comes from a variety of documents published by that society, particularly its annual prize lists.

4. Many articles have been written about Simonet and published in newspapers and horticultural magazines. Where I have used these they are cited. For biographical information, I have relied on interviews I conducted several years ago with Simonet's niece, Janine Dunn, in connection with another project. These took place on July 24 and September 25, 1995.

5. Archibald Oswald MacRae, *History of the Province of Alberta*, vol. 2 (Calgary: The Western Canada History Co., 1912), 848. MacRae saw Ramsay as someone whose business "contributed in large measure to the advancement of progress and prosperity" in Edmonton, "for no one can deny that the culture of beauty is refining to all connected with it"; McCracken, "Walter Ramsay Florist Ltd.," 2–8.

6. Gladys Reeves, letter to the *Edmonton Journal*, April 16, 1962, PR1974.173/33, PAA. Reeves was provoked to write this letter by reading the "Third Column," which had been published in the *Journal* on April 13, 1962.

7. M.A. Kostek, *A Century and Ten: The History of Edmonton Public Schools*, enlarged and updated edition (Edmonton: Edmonton Public Schools, 1992), 61.

8. McCracken, "Walter Ramsay Florist Ltd.," 18.

9. Advertisement, *EB*, December 21, 1905, 7.

10. "WM. [sic] Ramsay builds hothouse," *EB*, May 19, 1906, 1.

11. "Greenhouse opening," *EB*, October 15, 1906, 5.

12. McCracken, "Walter Ramsay Florist Ltd.," 80 and 105–12.

13. "Greenhouse opening," *EB*, October 15, 1906, 5.

14. "Greenhouse opening," *EB*, October 15, 1906, 5; McCracken, "Walter Ramsay Florist Ltd.," 68–69.

15. McCracken, "Walter Ramsay Florist Ltd.," 80–86.

16. McCracken, "Walter Ramsay Florist Ltd.," 84–88.

17. "Walter Ramsay," *EJ*, December 2, 1958, 4; "The Walter Ramsay house," *EJ*, March 22, 1960, 11.

18. McCracken suggests that the participation of Ramsay's son Donald in the business, which began after the end of the Second World War, was a significant factor in the longevity and vitality of the business.

19. Mary Shewchuk, while being interviewed by the author on August 22, September 20, and October 30, 2002, described her Ukrainian upbringing on a farm in northern Alberta and her parents' method of obtaining seed for cabbage.

20. The Seed Centre was opened around 1950 by Harry Wigelsworth, a market gardener in Edmonton and also a nephew to Alfred Pike. Wigelsworth worked for his uncle during the 1930s and 1940s until he decided to open the Seed Centre. Wigelsworth's son Richard joined the business in 1968. At the time of publication, the Seed Centre still operates as an Edmonton-based business. Richard Wigelsworth, interviews by author, January 15 and January 28, 2003.

21. "A.E. Potter, the city transfer company," *Town Topics*, Industrial and Investor's Number, 1913, 58, MS 325, Class 3, CEA.

22. Obituary of Alfred Pike, *EJ*, November 27, 1981, F4; *EJ*, November 30, 1981, clippings file for Alfred Pike and Pike & Co. Ltd., CEA.

23. H.W. Stiles, "The gardener," *EB*, May 14, 1932, 6, reported to the readers of his gardening column that his good friend "Mr. Pike, the seedsman," had supplied some bulbs for experiment and that the results had been very encouraging: "Had success with snowdrops, crocus, daffodils and narcissus."

24. This account is based on information supplied to the author by Walter Dorin (see note 3 above) and is consistent with photocopied documents supplied to the author by A.E. McKenzie Co., including the certificate of incorporation of the reorganized company issued by the Province of Alberta on October 28, 1953.

25. See Edmonton Horticultural Society prize lists, 1921–1936, MS 89, Class 4.3, Box 12, Files 1–16, CEA. The 1918 prize list, missing from the collection at the CEA but accessed by the author at the offices of the horticultural society, lists Alfred Pike as second vice-president, as well as being a member of both the show committee and the vacant lots garden committee.

26. The Edmonton Horticultural Society prize list for 1933 lists A. Pike & Co. as a donor. Class 303 in the program asked for a display of roses consisting of twenty blooms, each to be placed in receptacles provided by the society. The prize was to be twenty-five rose bushes supplied by Pike's company, MS 89, Box 12, File 13, CEA.

27. "Edmonton is rose city of west Canada," *EB*, April 27, 1935, 19.

28. "Roses of every color seen in beautiful display at Capitol," *EB*, July 24, 1934, 6.

29. Edmonton Horticultural Society Prize List, 1938, MS 89, Class 4.3, Box 12, File 18, CEA; John McLean, "Little man from England brought roses to city," *EB*, June 8, 1949, 3.

30. Georges Bugnet, "The Search for Total Hardiness," *American Rose Annual* (1941): 111–15.

31. Agriculture Canada established stations at Brandon, Manitoba, and Indian Head, Saskatchewan, in 1896, and in Alberta: Lethbridge, 1906; Lacombe, 1907; Fort Vermilion, 1908; and Beaverlodge, 1914–16. The Government of Alberta opened a research station in Brooks, Alberta, in 1935.
32. David Carpenter, "Nomme de Plume," in *Writing Home: Selected Essays* (Saskatoon: Fifth House Publishers, 1994), 72–89.
33. Lac Ste. Anne Historical Society, Archives Committee, account based on interview with Georges Bugnet, *West of the Fifth: A History of Lac Ste. Anne Municipality* (Edmonton: The Institute of Applied Art Ltd. Educational Publishers, 1959), 54.
34. Papen, *Georges Bugnet*, 26, describes Bugnet as having succumbed to propaganda distributed by Frank Oliver, Minister of the Interior, but if Bugnet saw these propaganda brochures in 1904, they must have been sent out under the auspices of Oliver's predecessor in the post, Sir Clifford Sifton. See Friesen, *The Canadian Prairies*, 249–50.
35. Papen, *Georges Bugnet*, 27; Don Thomas, "University went out of its way to honor author," *EJ*, June 5, 1978, B; Lac Ste. Anne Historical Society, *West of the Fifth*, 51–52.
36. Papen, *Georges Bugnet*, 18.
37. John S. Moir, "Petitot Émile," in *Dictionary of Canadian Biography*, vol. 14, University of Toronto/Université Laval, 2003–, accessed November 9, 2012, http://www.biographi.ca/en/bio/petitot_emile_14E.html.
38. Papen, *Georges Bugnet*, 18.
39. Lac Ste. Anne Historical Society, *West of the Fifth*, 51–52.
40. "French author develops hardy, thornless rose," *EJ*, November 13, 1948, 6.
41. Papen, *Georges Bugnet*, 27–32.
42. Papen, *Georges Bugnet*, 32–34.
43. Lac Ste. Anne Historical Society, *West of the Fifth*, 51–54.
44. According to Papen, it was "dans cette solitude, vaste et silencieuse, au milieu de cette grave dignité de la nature que, loin des vexations et des vaines agitations du monde, et face à la sérénité impassible de la forêt géante, Georges Bugnet comprendra au plus intime de son être la vérité de la liberté et de la vie humaine" ("in this vast, silent loneliness, surrounded by nature's gravity, removed from daily trials and tribulations, and opposed by unfeeling otherness of the forest, Georges Bugnet saw through to the heart of the truth about liberty and human life") (author's own translation), *Georges Bugnet*, 31.

45. Papen, *Georges Bugnet*, 34–35; "Four Alberta Pioneers," *Alberta Horticulturist* 6, no. 2 (June 1967): 1.
46. "French author develops hardy thornless rose," *EJ*, November 13, 1948, 6.
47. Janet Bliss, "Pioneer recalls confrontation with Aberhart," *EJ*, October 27, 1977, 25; Letter from R.J. Hilton, Professor of Horticulture, University of Alberta, to Georges Bugnet, October 19, 1955, MS 96-72, Box 1, Bruce Peel Special Collections Library, University of Alberta.
48. Carpenter, "Nomme de Plume," 87.
49. Obituary, *EJ*, January 13, 1981, C6.
50. Papen, *Georges Bugnet*, 37, writes that "à une âme chrétienne comme la sienne, l'horticulture lui permit, grâce au contact intime et constant avec les lois secrètes et grandioses de la nature, d'admirer la sagesse divine que s'y dévoilait."
51. "Four Alberta Pioneers," 1.
52. Lac Ste. Anne Historical Society, *West of the Fifth*, 53–54.
53. Lac Ste. Anne Historical Society, *West of the Fifth*, 53; W.H. Alderman, ed., *Development of Horticulture on the Northern Great Plains* (St. Paul, MN: The Great Plains Region American Society for Horticultural Science, 1962), 99–102, provides an excellent account of the career and accomplishments of Hansen.
54. Lac Ste. Anne Historical Society, *West of the Fifth*, 51–53; Papen, *Georges Bugnet*, 37–40.
55. The Regional Woody Plant Test Project, carried out by Alberta Agriculture and Rural Development since 1983, includes the Bugnet Scots pine in its trial results and was accessed November 12, 2012, at http://www1.agric.gov.ab.ca/$department/deptdocs.nsf/all/opp4077.
56. J. Horace McFarland, The American Rose Society, letters to Mr. Georges Bugnet, Lake Majeau, Alberta, Canada, April 22, 1940; November 8, 1940; November 30, 1940; January 21, 1941; February 11, 1941; February 28, 1941, MS 96-72, Box 1, Bruce Peel Special Collections Library, University of Alberta.
57. Papen, *Georges Bugnet*, 39; Paul G. Olsen, "The Roses of Georges Bugnet," *The Rosebank Letter* 34 (November 15, 2000): 1–3.
58. Bugnet, "The Search for Total Hardiness," 111–15.
59. Olsen, "The Roses of Georges Bugnet," 1–3.
60. Bugnet, "The Search for Total Hardiness," 111–15.
61. "Horticulturist Receives Gold Medal," *Alberta Horticulturist* 9, no. 2 (April 1971): 1.

62. Janine Dunn, interviews by author, July 24, 1995; September 25, 1995; June 13, 1998; "Grower of petunia seed has little competition," *EJ*, June 5, 1954, 36; "Robert Simonet: city petunia king," *EJ*, July 27, 1965, 17.
63. Fred Fellner, "Robert Simonet and the 'Rescued Lilies,'" North American Lily Society Yearbook, 1979, accessed March 5, 2012, http://www.manitobalilies.ca/Robert_Simonet.htm.
64. Dunn, interviews by author; "Robert Simonet," *Alberta Horticulturist* 5, no. 4 (December 1966): 4.
65. A subcommittee of the Alberta Horticultural Advisory Committee, headed by Roger Vick of the University of Alberta Devonian Botanic Garden, worked to rescue and preserve Simonet crosses. This effort was ultimately unsuccessful. Alberta Horticultural Advisory Committee Minutes, December 9, 1980, 1981, and 1982, SB 29 C2A14, Alberta Government Library, Great West Life Site.
66. Fellner, "Robert Simonet and the 'Rescued Lilies.'"
67. "City horticulturist wins Manitoba prize," *EJ*, December 17, 1959, 3. This article announced an award that was to be made at the Manitoba Horticultural Association convention in Winnipeg on February 11, 1960. The award was created in 1932 in honour of A.P. Stevenson, a horticulturist with the Morden Research Station whose specialty was hybridizing hardy fruit.
68. "Robert Simonet," 4.
69. Paul G. Olsen, "Robert Simonet's Roses," *The Rosebank Letter* 30 (March 15, 2000): 1–3.
70. "Robert Simonet," 4; "Trophies are presented at horticultural show," *EJ*, August 21, 1952, 5.
71. Ken Riske, conversation with author, July 2011; "Wanted, Plant Selections of Robert Simonet," *Alberta Horticulturist* 21, no. 2 (June 1983): 4; Olsen, "Robert Simonet's Roses," 1–3.
72. George Shewchuk, interviews by author, July 4 and 24, 2002; August 9 and 27, 2002; October 24, 2002; Bea Keeler, interviews by author, June 8, 17, and 25, 2002; Hugh Knowles, interviews by author, November 14, 21, and 28, 2002; January 16 and 30, 2003; February 13 and 21, 2003.
73. See the Alberta Agriculture and Rural Development website for the Alberta Agriculture Hall of Fame inductees at http://www1.agric.gov.ab.ca/$Department/deptdocs.nsf/all/info2060?opendocument.
74. "Horticulturist Receives Gold Medal," *Alberta Horticulturist* 9, no. 2 (April 1971): 1.
75. The Robert Simonet Graduate Scholarship is awarded annually to a full-time student in a graduate degree program in Agricultural, Food and Nutritional

Science (Horticulture) who has completed at least one year of course work and research towards an MSC or PHD degree. In addition, two travel bursaries may be applied for each year to graduate students wishing to attend scientific conferences.

76. H.T. Allen, Horticulturist, Canada Research Station, Lacombe, Alberta, addressing the Alberta Horticultural Association at the Brooks Horticultural Station on August 27, 1967, and reprinted in the *Alberta Horticulturist* 6, no. 3 (November 1967): 3–4; David Carpenter, introduction to *Writing Home: Selected Essays* (Saskatoon: Fifth House Publishers, 1994), x.

5 The Edmonton Horticultural Society: Working for the City Beautiful

1. Gladys Reeves, notes for a speech, n.d., PR1974.173/39a, PAA.
2. Frederick G. Todd, "Report [to the City of Edmonton] on Parks and Boulevards," April 1907, clippings file, CEA.
3. Secretary, Edmonton Horticultural Society, "Past, present and future," *Edmonton Town Topics Industrial Number*, 1913, 22 and 27, MS 325, Class B, CEA. This article, written just after the two horticultural societies, "following the lead given them by the two Cities," joined forces under the name "Edmonton Horticultural society," suggests that the Strathcona society had been formed eleven months prior to the Edmonton society, that is in December 1908 or January 1909.
4. "Plan to beautify Edmonton," *EB*, November 18, 1909, 10; "Around the city," *EB*, November 29, 1909, 8; "Edmonton Horticultural Society," *The Saturday News*, December 4, 1909, 11.
5. William H. Wilson, *The City Beautiful Movement* (Baltimore: Johns Hopkins University Press, 1989), 1.
6. Wilson, *The City Beautiful Movement*, 1.
7. Reeves was a professional photographer with an interest in horticulture and a commitment to the cause of city beautification. Cardy was employed by the City of Edmonton in the Electric Light Department; horticulture was a personal interest to which he devoted much time and creative energy. For Harcourt, Stowe, and Pike, horticulture was a career as well as a personal interest. All of these Edmontonians were active members of the Edmonton Horticultural Society.
8. Wilson, *The City Beautiful Movement*, 9, 18, 22–32.

9. Todd, "Report [to the City of Edmonton] on Parks and Boulevards," 3; Morell & Nichols, Landscape Architects, "A Report on City Planning," presented November 1912, clippings file, CEA.
10. Wilson, *The City Beautiful Movement*, 29 and 31.
11. Wilson, *The City Beautiful Movement*, 3.
12. James H. Marsh, ed., *The Canadian Encyclopedia*, 2nd ed. (Edmonton: Hurtig Publishers, 1988), s.v. "City Beautiful Movement," by Edwinna von Baeyer.
13. "The Edmonton Horticultural and Vacant Lots Garden Association Constitution and By-laws, 1920," PR1974.173/36, PAA; "The Edmonton Horticultural and Vacant Lots Garden Association, Constitution and By-laws, as revised and adopted, 1938," and "The Edmonton Horticultural Society, Articles of Incorporation and By-laws, April, 1973," MS 89, Class 1.1, Box 1, File 1, CEA; and Edmonton Horticultural Society By-laws, Revised November 29, 2004, author's copy. These four sets of bylaws are the ones referred to in the following text.
14. "Edmonton Horticultural Society," *The Saturday News*, December 4, 1909, 11.
15. "Vacant lots garden club is proposed for the city," *EB*, March 18, 1916, 11. See also, "Vacant Lots Garden Club," *EB*, March 25, 1916, 2, which describes the organizational meeting during which Harcourt was elected chairman of the new club with Mrs. Bishop, of the Local Council of Women, as vice-chairman. This article suggests that the new club "is evidently going to be a force in the city this summer."
16. The author participated as a member of the horticultural society's board of directors in discussions leading to the 2004 constitution and bylaws.
17. E.J. Stowe and W. Cardy, letter to Mayor D.K. Knott, February 15, 1932, RG 11, Series 3, Sub-series 3.1, Box 9, File 135, CEA. See also in this file an excerpt from the Commissioners' Report #15 to Council, dated February 22, 1932, and recommending against a takeover of the horticultural society by the City of Edmonton.
18. From 1920 until 1940, and again from 1958 to 1960, Henderson's Directories list an address for the Edmonton Horticultural Society in the Civic Block (Henderson's Directories, Reference Room, CEA), but it would seem likely that the society occupied this same address from 1918, the year it amalgamated with the Vacant Lots Garden Club and assumed responsibility for administering the vacant lots program for the city.
19. On January 19, 1940, Commissioner R.J. Gibb wrote to J. Martland, building inspector, "It is understood that the Horticultural Society will be

accommodated at the old Police Station," RG 11, Series 1, Sub-series 1.3, Box 3, File 47, CEA.

20. Correspondence relating to the collaboration between the city and the horticultural society on issues related to vacant lots from 1926 to 1967 can be found in RG 11, Series 3, Sub-series 3.1, Box 9, Files 133–40, CEA. This correspondence shows that the horticultural society provided a yearly report to the city commissioners highlighting the beautification projects it supported through revenues from vacant lots.

21. Wm. C. McMoore, Secretary-Treasurer, City of Edmonton, letter to the Secretary of the Edmonton Horticultural Society, March 20, 1912, stating that city council had authorized a three-hundred-dollar grant to the society, MS 209, File 141, CEA.

22. City Clerk, letter to City Commissioners, May 11, 1926, informing the commissioners about council's decision to award an extra fifty-dollar payment to the horticultural society; and G.E. Mantle, Secretary, letter to J. Hodgson, City Comptroller, May 13, 1926, authorizing him to pay fifty dollars to the horticultural society in addition to the regular two-hundred-dollar appropriation, RG 11, Series 1, Sub-series 1.4, Box 20, File 471, CEA.

23. City of Edmonton Financial Statements, 1912–1925, GP 464, CEA.

24. "Mayor presents prizes won by flower show exhibitors," *EJ*, August 22, 1929, 10.

25. "Edmonton Horticultural and Vacant Lots Garden Society [sic], Revenue from Vacant Lots, 1927–1931," undated, includes for each year the amount spent by the society on public service and the deficit on the operation of vacant lots, RG 11, Series 3, Sub-series 3.1, Box 9, File 135, CEA.

26. "President's Report," Annual prize list of the Edmonton Horticultural and Vacant Lots Garden Association, 1939, MS 89, Class 4.3, Box 12, File 19, CEA.

27. W.J. Cardy, President, Edmonton Horticultural and Vacant Lots Garden Association, letter to His Worship the Mayor, City of Edmonton, April 9, 1939; City Commissioners, letter to W.J. Cardy, October 16, 1940, thanking him for the fifty-dollar donation; G.C. Hunter, Secretary for the horticultural society, letter to City Commissioners, February 23, 1942, sending the donation towards caring for the cenotaph; City Commissioners, letter to City Comptroller, February 24, 1942, sending a fifty-dollar cheque "towards the cost of some City undertaking," RG 11, Series 3, Sub-series 3.1, Box 9, File 138, CEA.

28. President's reports, Annual prize lists of the Edmonton Horticultural Society, 1926–1933, MS 89, Class 4.3, Box 12, Files 5–13, CEA.

29. E.P. Williams, secretary Edmonton Horticultural and Vacant Lots Garden Society [sic], letter to the Mayor and Commissioners, City of Edmonton, March 1929; Mayor A.U.G. Bury, to E.P. Williams, secretary, March 22, 1929, RG 11, Series 3, Sub-series 3.1, Box 9, File 134, CEA.
30. President's report for 1932, Annual prize list of the Edmonton Horticultural Society, 1933, MS 89, Class 4.3, Box 12, File 13, CEA; George Buchanan, letter to Mayor Daniel Knott, June 16, 1932; Mayor Daniel Knott to A.W. Haddow, June 17, 1932, RG 11, Series 3, Sub-series 3.1, Box 9, File 135, CEA.
31. Annual prize lists for the Edmonton Horticultural Society, 1966 and 1967, MS 89, Class 4.3, Box 14, Files 44 and 45; "History of the Edmonton Horticultural Society," undated and author(s) not named, which takes the society up to the early 1990s, MS 89, Class 4.2, Box 11, File 1, CEA.
32. "History of the Edmonton Horticultural Society," MS 89, Class 4.2, Box 12, File 1, CEA.
33. See Kathryn Chase Merrett, "Public and Private Gardens for EHS Members, 1900–2000," and "The EHS Centennial Garden," in *A Century of Gardening in Edmonton*, ed. Maggie Easton et al. (Edmonton: Edmonton Horticultural Society, 2009), 41–43 and 50–51.
34. See note 33 above. The author chaired the committee for the Edmonton Horticultural Society that was responsible for planning this project with the City of Edmonton and ensuring that it was planted in 2007. Several years after the garden was planted, the City of Edmonton decided to route a light rapid transit (LRT) line through Henrietta Muir Edwards Park. At the time this book was being written, the horticultural society was working with the city on finding a suitable relocation site for its centennial garden.
35. Gladys Reeves, notes for a speech or a letter, n.d., PR1974.173/39b, PAA.
36. Gladys Reeves, "Summary of the Activities of the Edmonton Horticultural and Vacant Lots Garden Association," Annual prize list for the Edmonton Horticultural Society, 1925, MS 89, Class 4.3, Box 12, File 5, CEA; "Horticultural exhibition opens with good display," and "Toronto show is not as good," *EB*, August 27, 1924, 3.
37. Gladys Reeves, notes for a speech, n.d., PR1974.173/39a, PAA.
38. Gladys Reeves, notes for a speech to an unidentified group, n.d. [1926–1928], 1974.173/63, PAA.
39. Gladys Reeves, notes for a speech, n.d., PR1974.173/39a, PAA.
40. "Tree planting effort of woman to beautify western Canadian city," *Montreal Daily Star*, May 23, 1931, PR1965.22/1, PAA.

41. Gladys Reeves, letter to Ernest Brown, September 23, 1926, PR1965.22/4, PAA.
42. Commissioners' Report No. 16 to council, April 25, 1924, 6–7, RG 11.1, Class 1, File 18, CEA; "Clean-up campaign to buy 1300 trees to plant here," *EB*, April 19, 1924, 11; "Plant up week starts as clean up week ends," *EB*, May 5, 1924, 9.
43. See letter under the committee's letterhead, PR1974.173/59, PAA.
44. Gladys Reeves, notes for a speech, n.d., PR1974.173/39a, PAA.
45. "Plan to plant trees at Memorial grounds," *EB*, April 23, 1924, 1. Memorial Hall, predecessor to the various legion halls built in Edmonton as meeting places for veterans of the wars in which Canada has fought, was opened on a prestigious site on MacDonald Drive in April 1920. Designed by architect William G. Blakey, and built with funds raised partially by public subscription, it was intended to be both a memorial to soldiers who had fought in the First World War and as a home for those who returned in need of accommodation. Memorial Hall was demolished in either 1968 or 1969 to make way for the Alberta Government Telephone tower. See clippings files for Memorial Hall and for the Royal Canadian Legion, Montgomery Branch, CEA.
46. "Memorial tree planting ceremony this afternoon," *EB*, May 3, 1924, 11.
47. Gladys Reeves, notes for a speech, n.d., PR1974.173/39a, PAA.
48. Commissioners' Report No. 28, April 27, 1925, RG 11.1, Class 1, File 41, CEA.
49. Gladys Reeves, letter to the editor of the *Edmonton Journal* titled "Tree Planting Association replies," n.d. [spring 1928?], PR1974.173/33, PAA.
50. Gladys Reeves, letter to Ernest Brown, August 14, 1926, PR1965.22/4, PAA. Reeves frequently used abbreviations, such as the symbol "+" for the word "and." Where these have been used they have been edited to replace the symbol with the word.
51. Gladys Reeves, letter to Miss M.V. Johnson, *Winnipeg Free Press*, October 23, 1933, regarding a visit the latter had paid to Edmonton, PR1974.173/38, PAA.
52. Tom Raisbeck, "Gladys Reeves, 83, pioneer, noted photographer, dies," *EJ*, April 27, 1974, 39.
53. Obituary, *EJ*, April 12, 1960, 36; W.J. Cardy, obituary, *Alberta Horticulturist* 1, no. 1 (April 1962): 4.
54. W.J. Cardy, obituary.
55. Cardy won a first prize for his peonies as recorded in the annual show results, *EB*, July 18, 1921, 5; "Governor Brett opens flower show Wednesday," *EB*, August 16, 1922, 12; Bea Keeler, interviews by author, June 8, 17, and 25, 2002.
56. Cardy outlined the horticultural society's public service and horticultural outreach initiatives in the president's reports included in the horticultural

57. W.J. Cardy, letter to the City Commissioners' office, attention His Worship Mayor Clarke, August 1, 1936, RG 11, Series 3, Sub-series 3.1, Box 9, File 137, CEA.
58. President's reports 1936–1939, Annual prize lists of the Edmonton Horticultural Society, 1937–1940, MS 89, Class 4.3, Box 12, Files 17–20, CEA.
59. W.J. Cardy and H.C. Hunter, letter to the City Commissioners, March 8, 1943; and Mayor Fry, letter to W.J. Cardy, President, Edmonton Horticultural Society, November 1, 1943, in which the mayor asks Cardy to respond to A.M. Shaw of the Canadian Agricultural Supplies Board, RG 11, Series 3, Sub-series 3.1, Box 9, File 139, CEA.
60. The president and/or manager of the Edmonton Exhibition Association was listed in the horticultural society's annual prize lists as an honorary show patron from 1944 until 1955, just one indication of a relationship between the two organizations that continued for many years. See MS 89, Class 4.3, Boxes 13 and 14, Files 25–42, CEA. For a copy of the 1941 memorandum of agreement regarding the horticultural society's building of a storage facility at the exhibition grounds, see MS 89, Box 1, Class 1.2, File 6, CEA.
61. "Minutes of a meeting held in Calgary, August 26, 1950, to discuss the formation of a Provincial Association," "Minutes of a meeting held in Calgary, November 16, 1950 by the temporary committee to discuss the formation and organization of a Provincial Horticultural Association," and "Minutes of a meeting held July 3 and 4 [1951] at the Olds School of Agriculture to discuss the formation of a Provincial Horticultural Association," PR2003.0311/33, PAA. Minutes of the annual meeting of the Alberta Provincial Horticultural Association held on January 12, 1952, in Edmonton, reconfirmed the officers who had acted in their positions through the incorporation, including W.J. Cardy as president, PR2003.0311/34, PAA. The Alberta Horticultural Association was first incorporated in 1951 as the Alberta Provincial Horticultural Association, with the word "Provincial" being dropped in 1958. See G.M. Ramsay, "The Alberta Horticultural Association," *Alberta Horticulturist* 1, no. 1 (February 1962): 4.
62. In 1956, when the Edmonton Horticultural Society joined with the Alberta Horticultural Association to hold a joint show in August, Cardy's children donated a shield named after their father to be awarded to the show's grand aggregate winner. The Cardy Shield, which the family hoped would "remind everyone in the Province interested in Horticulture of the effort and

enthusiasm he has shown at all times towards encouraging and promoting Horticulture in this Province," continues to be awarded at provincial horticultural shows. See "Minutes of an Executive Meeting of the Alberta Horticultural Association," August 25, 1956, PR2003.0311/34, PAA.

63. "Best flower show in Edmonton's history is opening night verdict," EJ, August 17, 1927, 11–12; "Visiting poultrymen pay glowing tribute to Edmonton flowers," EJ, August 18, 1927, 12; "Attendance at flower show broke record," EJ, August 19, 1927, 13.

64. "Entries attain high standards," EJ, August 22, 1928, 9.

65. "History of the Edmonton Horticultural Society," n.d., MS 89, Class 4.2, Box 11, File 1, CEA.

66. P.D. McCalla, provincial horticulturist from 1949 until his retirement in 1981, worked closely with horticultural groups around the province of Alberta, especially the Alberta Horticultural Association that he had worked with Cardy to bring into being and which he continued to support both in an official and in a volunteer capacity. It is probably safe to assume that Cardy and McCalla shared a vision for horticulture in the province and believed that, through horticultural societies, amateurs and professionals could pool their resources for the common good.

67. See Government of Alberta, Department of Agriculture, *Annual Report*, 1956, 36, Alberta Government Library, Great West Life Site, S 135 A33. The Alberta Horticultural Advisory Board was to be composed of representatives from the Field Crops Branch, University of Alberta, Experimental Farms Service, the Science Service of the Canadian Department of Agriculture, and the Canadian Nurserymen's Association. It was charged with the responsibility of recommending to the minister "regarding all matters pertaining to horticulture."

68. Annual prize list of the Edmonton Horticultural Society, 1951, MS 89, Class 4.3, Box 13, File 31, CEA.

69. "Report: Recommendations for the future direction of the Alberta Horticultural Association," and "Alberta Horticultural Association, statement of receipts and disbursements for the nine months ended September 30, 1992," Alberta Horticultural Association fonds, 2003.0311/33, PAA.

70. For information about Communities in Bloom, see http://www.communitiesinbloom.ca.

71. John Helder, interviews by author, November 18, 2002; April 9, 2003; February 27, 2004; July 24, 2012. As principal of horticulture for the City of Edmonton

from 1995 to 2013, Helder was responsible for entering the city in the first annual Communities in Bloom event, a tradition the city has maintained.

72. The annual show tradition was revived in 2009 as part of the horticultural society's centennial celebrations. It was restaged in 2010 and 2011 and then discontinued again.

73. Rodney M.B. Al, "The EHS's Current Decade: Edmonton's Front Yards in Bloom, 1999–2008," in *A Century of Gardening in Edmonton*, ed. Maggie Easton et al. (Edmonton: Edmonton Horticultural Society, 2009), 55–57.

6 Waste Places: Vacant Lot Gardening in Edmonton

1. To view the film, see the website of the National Film Board, accessed February 27, 2012, at http://www.nfb.ca/film/he_plants_for_victory.
2. J.E. Pember, "Vacant lots gardens make waste places in the City of Edmonton veritably blossom as the rose," *EB*, July 22, 1918, 5.
3. Laura J. Lawson, *City Bountiful: A Century of Community Gardening in America* (Berkeley: University of California Press, 2005), 3.
4. See the website of the Ontario Horticultural Association, accessed December 1, 2012, http://www.gardenontario.org. Select "about OHA," then click the link "OHA: From Then to Now."
5. Horticulture clippings file, GA.
6. "Vacant lots garden club is proposed for the city," *EB*, March 18, 1916, 11.
7. "Vacant Lots Garden Club," *EB*, March 25, 1916, 2; Walter H. Johns, *A History of the University of Alberta, 1908–1969* (Edmonton: University of Alberta Press, 1981), 30–31, 49.
8. The Honourable Dr. Robert G. Brett was Alberta's Lieutenant-Governor from 1915 to 1925. Throughout his tenure as Lieutenant-Governor, he promoted gardening and championed the work of the Edmonton Horticultural Society, especially in speeches given at the society's annual horticultural shows. See, for instance, "'Make a garden' says lieutenant governor; help your community and also find health in doing so," *EB*, August 17, 1922, 1 and 3; "Horticultural exhibition opens with good display," *EB*, August 27, 1924, 3.
9. "Vacant lots garden club is proposed for the city," *EB*, March 18, 1916, 11.
10. Pember, "Vacant lots gardens make waste places"; George Harcourt, "Report of the Chairman of the Vacant Lots Committee," MS 89, Box 9, File 1, CEA.
11. Alderman, *The Development of Horticulture on the Northern Great Plains*, 176.
12. Roger Vick, "George Harcourt, Horticulturist," *Alberta History* 39, no. 3 (Summer 1991): 21.

13. "Annual Report of the Department of Agriculture of the Province of Alberta from the first of September, 1905, to the thirty-first of December, 1906" (Edmonton: Government Printer, 1907), 160–61, accessed December 3, 2012, http://archive.org/details/annualreport1905albeuoft.
14. The Harcourt apple "originated from an open pollinated seedling taken from an apple tree in Gleichen, Alberta, in 1930. It was selected by the late Prof. George Harcourt and listed as strain UA2 in 1950. In 1955 it was named the Harcourt variety." See "Apple Variety Being Developed," *Alberta Horticulturist* 5, no. 4 (December 1966): 3–4.
15. George Harcourt, "Report of the Chairman of the Vacant Lots Committee," MS 89, Box 9, File 1, CEA.
16. "Report of the Canada Food Board: February 11–December 31" (Ottawa: 1918), 1–5, accessed March 1, 2012, http://www.archive.org/details/cu31924013848696.
17. "Report of the Canada Food Board," 66–67.
18. Edmonton Horticultural Society Prize List, 1918. This prize list was viewed by the author in the offices of the Edmonton Horticultural Society and some pages were photocopied. It has since disappeared and cannot be located.
19. George Harcourt, "Report of the Chairman of the Vacant Lots Committee," MS 89, Box 9, File 1, CEA.
20. "Some 7000 vacant lots under local cultivation should produce $100,000," *EJ*, June 26, 1918, 2.
21. Pember, "Vacant lots gardens make waste places."
22. M.J. O'Farrell, Superintendent, Edmonton Horticultural and Vacant Lots Garden Association, letter to owner of Lot 10, Block 10, Acre Lot 1, Westgrove, (Mr. Arthur Buswell), n.d. (received June 16, 1920), MS 319, Class 5, File 35, CEA.
23. "Plant 35 acres relief gardens," *EJ*, May 20, 1933, 14. According to David Evans, who researched and wrote for the City of Edmonton, *The City of Edmonton: History of the City Council, 1892–1977*, Mayor Daniel Knott appointed H.F. McKee as superintendent of the special relief department immediately after the former's election as mayor in November 1932 (39). The special relief department may have begun to function as part of the relief department some time in 1931.
24. Edmonton Horticultural Society Prize List, 1930, MS 89, Class 4.3, Box 12, File 10, CEA; G.S. Botzow, Secretary-Treasurer Edmonton Horticultural and Vacant Lots Garden Association, letter to the City Commissioners, October 31, 1930, RG 11, Series 3, Sub-series 3.1, Box 9, File 134; E.J. Stowe, to City Commissioners, October 29, 1931, RG 11, Series 3, Sub-series 3.1, Box 9, File 135; W.A. Stowe

(secretary) to City Commissioners, November 23, 1935, and W.A. Stowe to City Commissioners, December 21, 1936, RG 11, Series 3, Sub-series 3.1, Box 9, File 137; Edmonton Horticultural Society to City Commissioners, November 9, 1939, RG 11, Series 3, Sub-series 3.1, Box 9, File 138, CEA.

25. "New cuts range from 1 to 16 p.c.," *EB*, February 14, 1935, 9. The mayor was quoted in this article as having said about McKee, "He's more humane than most of us think."

26. For example, relief gardeners were described as the "greatest offenders" as regards weed management. See E. Jarron, Secretary, Edmonton Horticultural Society, letter to City Commissioners, December 3, 1934, RG 11, Series 3, Sub-series 3.1, Box 9, File 137, CEA. The relatively meager participation by relief gardeners in the classes established particularly for them, and the indifferent results were not emphasized by McKee in his reports to the commissioners. For example, see H.F. McKee, Manager, Special Relief Department, letter to City Commissioners, May 15, 1937, RG 11, Series 7, Sub-series 7.4, Box 5, File 53, CEA.

27. Edmonton Horticultural Society to H.F. McKee, August 1, 1933, RG 11, Series 3, Sub-series 3.1, File 136, CEA.

28. H.F. McKee, letter to City Commissioners, May 15, 1937, and Commissioner Gibb, letter to H.F. McKee, May 19, 1937, RG 11, Series 7, Sub-series 7.4, Box 5, File 53, CEA.

29. H.F. McKee, Manager, Special Relief Department, letter to City Commissioners, August 14, 1937, and "Report based on judges' comments in unemployed garden competition," n.d. [1937], RG 11, Series 7, Sub-series 7.4, File 54, CEA.

30. "Report of Judging in the Relief Departments [sic] Flower and Vegetable Competition," RG 11, Series 7, Sub-series 7.4, Box 5, File 58, CEA.

31. F.W. Charlesworth, letter to Mayor Douglas, April 10, 1930; Mrs. F.W. Charlesworth, letter to Mayor Douglas, April 11, 1930; Mayor J.M. Douglas, letter to F.W. Charlesworth, April 14, 1930, RG 11, Series 3, Sub-series 3.1, Box 9, File 134, CEA.

32. Antoinette Grenier, interviews by author, August 3 and 24, 2004; November 3, 2004; February 9, 2005.

33. Evans, *The War on Weeds*, 74.

34. "The Russian thistle," *EB*, August 20, 1894, 2.

35. "For Calgary fair," *EB*, September 14, 1920, 2.

36. "Destruction of noxious weeds," *EB*, July 8, 1901, 2.

37. "Local," *EB*, August 2, 1901, 6.

38. Evans, *The War on Weeds*, 125.
39. A.W. Haddow, City Engineer, letter to Edmonton Horticultural Society, November 29, 1934, RG 11, Series 3, Sub-series 3.1, Box 9, File 136.
40. William Cardy and E.P. Williams, letter to City Commissioners, November 4, 1926; S.B. Ferris, Superintendent, Land Department, report to the City Commissioners, November 25, 1926; G.E. Mantle, Secretary to the Mayor, letter to Mr. Cardy, President, Horticultural and Vacant Lots Garden Association, December 20, 1926, RG 11, Series 3, Sub-series 3.1, Box 9, File 133, CEA.
41. "Victory food garden guide," *EB*, April 24, 1943, 20.
42. W.T.H., "Hard headed business man succumbs to garden poetry," *EB*, May 19, 1943, 9.
43. "The victory gardens," *EB*, July 29, 1943, 4.
44. Kay Ford, "City backyard transformed into fairylike playground," *EB*, July 20, 1944, 11.
45. W.J. Cardy, President, Edmonton Horticultural Society, letter to City Commissioners, March 8, 1943, RG 11, Series 3, Sub-series 3.1, Box 9, File 139, CEA.
46. H.C. Hunter, Secretary, Edmonton Horticultural Society, letter to A.M. Shaw, Chairman, Agricultural Supplies Board, Ottawa, November 10, 1943, RG 11, Series 3, Sub-series 3.1, Box 9, File 139, CEA. A brief article titled "City Farmers," which appeared on page 14 of the June 1944 issue of *The Country Guide*, sought to inform farmers "who are not inclined to attach much importance to the pokey little vegetable gardens on city lots, to know that wartime gardens in urban centres of more than 1,000 population, produced last year about 115 million pounds of vegetables." The *Guide* estimated that just over 209,000 wartime gardens had averaged 550 pounds of vegetables each. The most popular crop was potatoes, followed by tomatoes, carrots, and beets.
47. W.J. Cardy, President, Edmonton Horticultural Society, letter to City Commissioners, March 8, 1943, RG 11, Series 3, Sub-series 3.1, Box 9, File 139, CEA.
48. Lawrence Herzog, "Westmount mall at 50," *Edmonton Real Estate Weekly* 23, no. 35 (September 1, 2005), accessed May 22, 2013, http://www.rewedmonton.ca/content_view_rew?CONTENT_ID=1147.
49. Commissioners' Report No. 4, December 8, 1947, RG 11, Series 3, Sub-series 3.1, Box 9, File 140, CEA.
50. Walter Dorin and Ethel Dorin, interviews by author, November 26, 2003; February 2, 2004; March 4, 2004.

51. Bea Keeler, interviews by author, June 8, 17, and 25, 2002.
52. A. Young, Secretary-Treasurer, Edmonton Horticultural Society, letter to Civic Commissioners, November 29, 1956, RG 11, Series 3, Sub-series 3.1, Box 9, File 140, CEA; A. Young, Secretary-Treasurer, Edmonton Horticultural Society, letter to City Commissioners, January 13, 1960, RG 11, Series 3, Sub-series 3.1, Box 9, File 140, CEA; "History of the Edmonton Horticultural Society," n.d., MS 89, Class 4, Sub-class 4.2, Box 12, File 1, CEA.
53. John Helder, interviews by author, November 18, 2002; April 9, 2003; February 27, 2004; July 24, 2012.
54. Susan Penstone, interview by author, as revised with input from Penstone, April 12, 2005.
55. Sustainable Food Edmonton concerns itself with issues and initiatives involving food literacy and food security. It operates several programs, including Community Gardens, Yard Share, Little Green Thumbs, and Urban Ag High, and it is the former operator of a popular program known as City Farm. It came into being in 1989 as the Personal Community Support Organization but was renamed and redefined in 2011 when it was officially registered under its current name. Information about Sustainable Food Edmonton and its programs can be found at http://sustainablefoodedmonton.org.

7   Edmonton, the Rose City

1. In 1976, G.W. Shewchuk wrote a seventeen-page pamphlet on growing roses in northern Alberta. This was followed by Shewchuk's self-published *Growing Roses in the Prairie Provinces: Proven Techniques* (1981), and G.W. Shewchuk, *Rose Gardening on the Prairies* (Edmonton: University of Alberta, Faculty of Extension, 1988). The 1988 book, revised and expanded, was published as G.W. Shewchuk, *Roses: A Gardener's Guide for the Plains and Prairies* (Edmonton: University of Alberta, Faculty of Extension, 1999; revised edition). See also, Harry McGee, "Geo. Shewchuk's Magnum Opus," *The Rosebank Letter*, no. 29 (January 15, 2000): 8; George Shewchuk, "1998 Review of New Roses," *The Rosebank Letter*, no. 23 (January 15, 1999): 2, 3, 6.
2. See Royer and Dickinson, *Plants of Alberta*, 60.
3. The three wild rose species commonly found in the Edmonton region are *Rosa acicularis* or prickly rose, *Rosa arkansana* or low prairie rose, and *Rosa woodsii* or Wood's rose. See Royer and Dickinson, *Plants of Alberta*, 60–61.
4. Royer and Dickinson, *Plants of Alberta*, 60; information about the prickly rose accessed March 5, 2012, at http://www.albertarose.com.

5. The *Edmonton Bulletin* reported on June 11, 1887, that "[w]ild roses are in bloom." On September 19, 1891, it boasted that "[w]ild roses are still in bloom," and on June 23, 1892, wild roses were reported to be "in bloom abundantly." These and other references to fruits and flowers were made in the spirit of boosterism, and they combined to convey the impression that Edmonton and its surrounding district were fertile, naturally productive, and attractive.
6. "Local," *EB*, August 11, 1888, 1.
7. P.D. McCalla, "The Wild Rose," *Alberta Horticulturist* 5, no. 3 (October 1966): 3–4.
8. *EB*, March 7, 1885, 1.
9. "Local," *EB*, August 25, 1888, 1.
10. See the "Local" column in the *Edmonton Bulletin* on July 5, 1890, 1; August 2, 1890, 1; August 23, 1890, 1; July 25, 1891, 1; July 31, 1893, 1; July 1, 1895, 1; and July 15, 1895, 1.
11. "Donald Ross' greenhouses," *EB*, January 3, 1903, 1.
12. "Past, present and future," *Edmonton Town Topics*, Industrial and Investor's Number, 1913, 22 and 27, MS 325, Class 3, CEA.
13. "Strathcona flower show a big success," *EB*, August 10, 1911, 2.
14. Advertisement for rose show at the Capitol Theatre, *EB*, July 17, 1933, 9.
15. "Shrubs and vines will make great difference in garden," *EB*, May 5, 1924, 2.
16. "Best flower show in Edmonton's history is opening night verdict," *EJ*, August 17, 1927, 11–12.
17. "Flower show results," *EB*, August 23, 1928, 3; H.A. Holland, "Start right if you want good roses," *EB*, April 13, 1929, 25.
18. See the following columns by H.W. Stiles in the *Edmonton Bulletin*: April 16, 1932, 17; April 30, 1932, 3; May 7, 1932, 7; June 18, 1932, 5; and June 25, 1932, 5. See also, "Edmonton gardens at best; fine profusion of blooms," *EJ*, August 9, 1932, 2.
19. "Prominent visitors coming for annual show of roses: one man's faith pays off," *EB*, July 17, 1950, 9; "Edmonton's 'Mr. Theatre,' Walter Wilson, retiring," *EJ*, January 13, 1954, 10.
20. Barbara Paterson, telephone conversation with author, April 15, 2003; advertisement, *EB*, July 17, 1933, 9.
21. "Cap. Theatre crowded for rose display," *EB*, July 25, 1933, 5.
22. "Roses of every color seen in beautiful display at Capitol," *EB*, July 24, 1934, 6.
23. The University of Alberta Devonian Botanic Garden collection of seed catalogues.

24. "Capitol Theatre's 9th annual rose show to be held on Monday, Tuesday," *EB*, July 26, 1941, 16; "Capitol Theatre's 10th annual rose show to be held on Monday, Tuesday," *EB*, July 25, 1942, 6; full-page advertisement, *EB*, Saturday, July 22, 1944, 16.
25. "Bulletin challenge cup for best rose bloom in show won by James Lynch," *EB*, July 29, 1941, 9 and 16.
26. "Capitol Theatre's 10th annual rose show to be held on Monday, Tuesday," *EB*, July 25, 1942, 6; "10th annual rose show outstanding success," *EB*, July 28, 1942, 16.
27. "Capitol Theatre's 11th annual rose show to be held on Monday, Tuesday," *EB*, July 24, 1943, 14; "Multitude of lovely blooms at rose show," *EB*, July 26, 1943, 9; "Amateur rose grower wins Bulletin trophy for show's best bloom," *EB*, July 27, 1943, 9 and 16.
28. Advertisement for rose show, *EB*, July 15, 1950, 9; "Prominent visitors coming for annual show of roses: one man's faith pays off," *EB*, July 17, 1950, 9.
29. "Rose show with best," *EB*, July 25, 1950, 13; "Peace blossom named winning rose," *EJ*, July 24, 1951, 12. The 'Peace' rose was bred in France in 1935 and was introduced there in 1945 as 'Madame A. Meilland'. The same year it was introduced in the United States by Conard-Pyle (Star Roses) as 'Peace'. See HelpMeFind, a web-based reference work specializing in information about roses, peonies, and clematis, accessed May 2, 2012, at http://www.helpmefind.com/rose/l.php?l=2.2203.0.
30. "City woman wins five cups at flower show," *EB*, August 17, 1950, 11.
31. Dora Davies, "Edmonton's flower queen," *Canadian Homes and Gardens*, May 1957, 36–37 and 67–68, MS 226, Class 2, File 1, CEA.
32. Ernest Hodgson, handwritten notes, MS 226.1, File 1, CEA.
33. Dora Davies, "Edmonton's flower queen," *Canadian Horticulture and Home Magazine*, July–August 1943, 36–37, 67–68; Hilda May McAfee, "Starting from scratch, by a woman gardener," *Canadian Horticulture and Home Magazine*, February 1944, 32–33, MS 226, Class 2, File 1, CEA.
34. "Peace blossom named winning rose," *EJ*, July 24, 1951, 12.
35. University of Alberta Devonian Botanic Garden collection of seed catalogues.
36. April 20, 1954, is the postmark on a card sent to G. Reeves, 7832 Jasper Avenue, advertising an Edmonton Horticultural Society meeting for April 22nd at 8.00 P.M. in the North Side Library, PR1974.173/38, PAA. Roses were the favourite flower of Hendrik Arends, who came to Canada some time in the early 1920s with his wife, Cornelia, and who founded H. Arends' Rose Gardens

in the early 1930s, just as the campaign to make Edmonton a rose city got underway. According to Hendrik's daughter Corrie, who was interviewed by the author in February and in April 2005, her father's training in horticulture, all gained in Holland, went beyond practical skills to include a thorough grounding in botany. Roses were his particular passion. Arends imported stock directly from Holland and grafted hybrid teas onto hardy rootstock himself before selling them to customers.

37. "Chief provincial 'gardener' retires after 39 years," *EJ*, August 21, 1952, 5.
38. "Theatre-managing pioneer, 92, dies," *EJ*, March 27, 1969, 22.
39. Olsen, "The Roses of Georges Bugnet," 1–3.
40. Claude Roberto, interview by author, July 2003; Bugnet, "The Search for Total Hardiness," 111–15.
41. H.H. Marshall, "Prairie Hardy Roses," *Alberta Horticulturist* 18, no. 3 (September 1980): 3; Paul Olsen, "Forty-five Years of Parkland Roses," in *The 2008 Prairie Garden Annual*, ed. Richard Denesiuk (Winnipeg: The Prairie Garden Committee, 2007), 23–25; Bill Redekop, "Bloom off rose for Morden breeding program," *Winnipeg Free Press*, August 7, 2010; Peter Harris and André Imbeault, "The Mysterious Parentage of 'Thérèse Bugnet,'" *Roses-Canada*, no. 55 (July 2011): 3–6.
42. J. Horace McFarland, Editor, The American Rose Society, letter to Mr. Georges Bugnet, April 22, 1940, MS 96-72, Box 1, Bruce Peel Special Collections Library, University of Alberta; Olsen, "Robert Simonet's Roses," 1–3.
43. Arnold F. Pittao and Brian J. Porter, "Parkland and Explorer Roses Cultivar Data," in *The 2008 Prairie Garden Annual*, ed. Richard Denesiuk (Winnipeg: The Prairie Garden Committee, 2007), 30–37; Alberta Horticultural Advisory Committee, *Alberta Horticultural Guide* (Edmonton: Alberta Agriculture, n.d. [1978?]).
44. George Shewchuk, interviews by author, July 4 and 24, 2002; August 9 and 27, 2002; October 24, 2002; Harry McGee, "Octogenarian Honoured: a tribute to George Shewchuk," *The Rosebank Letter*, no. 4 (March 14, 1996): 2; McGee, "Geo. Shewchuk's Magnum Opus," 8.
45. Heiko Lotzgeselle, interview by author, May 7, 2012. The account of Lotzgeselle's rose garden is based on this interview.
46. Edgar Toop, "The Marigold: Edmonton's Floral Emblem," in *A Century of Gardening in Edmonton*, ed. Maggie Easton et al. (Edmonton: Edmonton Horticultural Society, 2009), 36–37.

8   The Invisible Tapestry: Remembering Edmonton's Chinese Gardeners

1. Information about the Dr. Sun Yat-Sen garden accessed September 14, 2012, at http://vancouverchinesegarden.com. The Kurimoto Garden was officially opened on September 7, 1990, by then Lieutenant-Governor Helen Hunley, who referred to it as a "peace place." See "A Peace Place," *New Trail* (Winter 1990): 24–25.

2. George Ng, interview by author, March 22, 2010. Ng joined the Chinese Garden Society in 2000 when it was still an informally constituted group. His first contribution was to help the group obtain status as a registered charity, a status it obtained in September 2000, and, two months later, to have it incorporated under the Alberta Societies Act as the Edmonton Chinese Garden Society. Ng became co-chair of the society in 2003. He was instrumental in the group's fundraising activities, including the obtaining of a $250,000 grant from the Infrastructure Canada-Alberta Program (ICAP) and four substantial grants from the Alberta Community Facilities Enhancement Program, three of $125,000 and another of $90,000.

3. The idea to create a formal, enclosed Chinese garden resurfaced in 2011 when the Chinatown and Little Italy Business Association put forward the idea to create one in a redevelopment of a park at the corner of 95th Street and 105th Avenue, in the heart of Chinatown. Paula Simons, a columnist for the *Edmonton Journal*, wrote a column describing the proposal that appeared in the *EJ*, January 6, 2011.

4. George Ng, "Edmonton Chinese Garden Society Business Plan," May 21, 2003 (Edmonton Chinese Garden Society). A copy of this report was given to the author by Mr. Ng.

5. I have not adopted a single convention for recording Chinese names. The Chinese convention puts the clan or family name first, followed by a shared generation name and, finally, a given name. Gradually, Chinese in Canada have adopted the Canadian convention for listing names, putting their first or given name before the family name. First-generation Chinese often stuck with the name order they had grown up with. Subsequent generations generally adopted the Canadian convention. I have followed the order of names as they appeared in directories or in correspondence or, in the case of Bark Ging and Young See Wong, as they are represented by family members. Marty Chan, "Chinese Garden Reflection," *Legacy*, (Summer 2009): 8–10. Subsequent references in this essay to Chan's reflections are to this article. Helen Gorman Cashman, "Sam Sing is friend and advisor to Edmonton Chinese," *EJ*, August

14, 1935, 17. Subsequent references to Cashman's work in this essay are to this newspaper article.

6. Obituary, Lock Chin, *EJ*, September 5, 1942, 20.

7. The City of Edmonton signed a ten-year lease with the Kinsmen Club that commenced on April 1, 1953, and was to end on March 31, 1963. The Kinsmen Club paid one dollar a year to the City of Edmonton for the privilege of developing an area that extended from the riverbank on the south to Walterdale Hill Road on the north and from 107th Street on the east to the High Level Bridge on the west. In following years, this lease was renewed and other leases were signed, enabling the Kinsmen Club to further develop the area as a sports complex, the facilities completed being turned over to the City of Edmonton to operate. A copy of the 1953 contract is in the Parks and Recreation clippings file for Kinsmen Park, CEA. In 1966, the *Edmonton Journal* carried an article suggesting that long-time residents of the area were not happy with the naming of the area as Kinsmen Park. A Mr. Roy Devore was the chief spokesperson for the move to ensure that the park became known as "Walterdale Park." See "Kinsmen? Walterdale? Park name in dispute," *EJ*, November 26, 1966, 26. Mr. Devore's "The Story of Walterdale: Historic Ground," n.d., is in the Gladys Reeves fonds, PR1965.22/50, PAA.

8. A copy of an aerial photograph that used to hang in the Pro Shop of Victoria Golf Course (it now hangs in the golf course's maintenance building) was given to the author during an interview on August 1, 2002, with Joe Craven, operations supervisor for the City of Edmonton; Kevin Hogan, Victoria Golf Course golf pro; and John Aiken, who worked there as a golf pro from 1950 to 2000. Aiken believed the photograph had been taken by photographer Alfred Blyth, but the only name on the print is Airline Photos. It looks as though it is a black and white photograph that has been hand-tinted for artistic effect.

9. Allan Shute and Margaret Fortier, *Riverdale: From Fraser Flats to Edmonton Oasis* (Edmonton: Tree Frog Press, 1992), 106–07 and 209.

10. Ging Wei Wong and Fook Kwan Wong, interviews by author (in person and via e-mail), 2003–2012.

11. Tony Cashman, "High Level the bridge that refuses to fall," *EJ*, August 24, 2012, noted that "The Royal Glenora Club sits on the site of the former Chinese market garden, a hiking trail on the route of the Edmonton Yukon and Pacific Railway," but he does not name the gardener. When Ken Tingley and Paul Bunner wrote *Heart of the City: A History of Cloverdale from Gallagher Flats to Village in the Park* (Edmonton: Cloverdale Community League, 2005), they wrote that

their interviewees "had particularly strong-and fond-memories of a Chinese man named 'Charlie' who operated a market garden on what is now the Muttart Conservatory grounds." None of the interviewees knew the gardener's last name, although he was "greatly admired for his kindly, generous ways" (96).
12. Kathryn Chase Merrett, *A History of the Edmonton City Market, 1900–2000: Urban Values and Urban Culture* (Calgary: University of Calgary Press, 2001), 71.
13. "Chief provincial gardener retires, after 39 years," EJ, August 21, 1952, 5.
14. The portion of Grand Trunk Park formerly occupied by the Wong family was known for years as Onion Park, so called because of the onions that persisted after the market gardening ended.
15. Eric Chen and Ruby Law, interview by author, February 11 and March 22, 2008. Chen and Law are the owners of the market gardening business Peas on Earth.
16. Francis Ng, interview by author, April 14, 2010.
17. Maggie Keswick, *The Chinese Garden: History, Art and Architecture*, revised by Alison Hardie (London: Frances Lincoln, 2003), 10.
18. The design of the Edmonton Chinese Garden was a team effort that involved bringing to Edmonton experienced garden designers from China. The Edmonton team, with Patrick Butler as landscape architect and prime consultant and Francis Ng as architect, worked through a translator with the Chinese team before proceeding to produce a final design and contract documents.
19. Francis Ng, "Edmonton Chinese Garden," undated and unpublished document prepared by Francis Ng to explain the inspiration and rationale behind the building of the Edmonton Chinese Garden and to describe its basic elements.

9   Citizen Gardeners

1. The language used in the 1925 advertisements for the city's annual cleanup campaign is moralistic in the extreme, equating dirt with evil and cleanliness with good. See, EJ, May 8, 1925, 8; EB, May 16, 1925, 8–9.
2. With the exception of the Edmonton Horticultural Society, the organized horticultural groups to which Reeves devoted so much of her time and energy seem to have been relatively short-lived. Judging from the notes she kept for her speeches, the Edmonton Tree Planting Committee was informally organized by a group of individuals from the horticultural society and from Edmonton's community leagues around a shared commitment to city beautification. The tree-planting committee remained active in Edmonton for many years, from about 1923 until some time in the late 1930s. The Alberta Horticultural and Forestry Association to which Reeves referred in notes for

her many speeches on city beautification, and which she once described as an umbrella organization whose goal it was to support the creation of horticultural and tree-planting groups across the province of Alberta, appears not to have lasted a long time; the perceived need among horticulturists for such an organization was eventually realized by the formation of the Alberta Horticultural Association in 1950–1951. See notes for Reeves's speeches in PR1974.173/39a, b, and c, PAA.

3. Reeves used the word "upbuilding" in a letter to the editor of the *Edmonton Journal* in which she outlined the contributions her father had made to his adopted home. Gladys Reeves, letter to Mr. Wallace, Managing Editor of the *Edmonton Journal*, n.d., PR1974.173/33, PAA.

4. Gladys Reeves, notes for speeches, n.d., PR1974.173/39a, PAA. Reeves used many abbreviations in her notes. Some quotations have been minimally edited to replace abbreviations with words.

5. The Keomi Club of Edmonton existed to further the cause of City Beautiful. One of its annual projects was to sponsor a competition among schools, a competition that was judged by Reeves and fellow horticultural society member George Buchanan. See notes for speeches referring to these competitions and to the Keomi Club and some of the reports on the competitions signed by Reeves and Buchanan in PR1974.173/63, PAA; "Children's Vehicles Provide Pretty Sight And Good Contest," *EB*, August 28, 1924, 6.

6. "All Attendance Records at Flower Show Broken," *EB*, August 19, 1926, 1; Gladys Reeves, letter to Ernest Brown, August 18, 1926, PR1965.22/4, PAA.

7. Gladys Reeves, notes for a speech, n.d., PR1974.173/39a, PAA.

8. "Plan to make city of blue and gold," *EB*, August 17, 1932, 6; Zoe Pauline Trotter, "Beautiful Homes and Gardens of Edmonton," PR1974.173/44, PAA; J.J. Meredith, Manager, Chateau Lake Louise, letter to Miss Gladys Reeves, August 29, 1932, PR1974.173/47.

9. George Harcourt, "Beautifying the Home Grounds," *The Press Bulletin*, issued by the Department of Extension of the University of Alberta, 15, no. 5 (May 9, 1930): 1–8, UAA, Accession No. 72-129:1–9; Reeves refers to Harcourt as the source of photos in notes for speeches on city beautification. See PR1974.173/39a and 39b, PAA.

10. "Who's who in the news," *EB*, May 3, 1930, 1.

11. On September 4, 2003, the author met with members of the Hole family, including Lois Hole, about their family gardening business. This interview was followed over the next few months by interviews with a number of the staff

at Hole's and by e-mail correspondence between the author and both Bill and Jim Hole. In April 2004, a profile of the business was assembled and approved by the family. The account of Lois Hole in this chapter is based partly on this interview but more heavily on her autobiographical book, *I'll Never Marry a Farmer: Lois Hole on Life, Learning & Vegetable Gardening* (St. Albert, AB: Hole's Publishing, 1998).

12. Lois Hole, *Lois Hole's Tomato Favorites* (Edmonton: Lone Pine Publishing, 1996), 8.
13. Hole, *I'll Never Marry a Farmer*, 11, 62–63.
14. Hole, *I'll Never Marry a Farmer*, 117.
15. Hole, *I'll Never Marry a Farmer*, 19.
16. Hole, *I'll Never Marry a Farmer*, 150.
17. Hole, *I'll Never Marry a Farmer*, 152–53.
18. Hole, *I'll Never Marry a Farmer*, 34.
19. Mark Kingwell, *The World We Want: Virtue, Vice, and the Good Citizen* (Toronto: Penguin Books, 2000), 5.
20. John Helder, interviews by author, November 18, 2002; April 9, 2003; February 27, 2004; July 24, 2012. Biographical information in the account that follows is drawn from these interviews.
21. For an account of the community garden project in Mill Woods, see "Waste Places: Vacant Lot Gardening in Edmonton" in this book.
22. For a fuller account of the Communities in Bloom and Front Yards in Bloom programs, see "The Edmonton Horticultural Society, Working for the City Beautiful" in this book. Citizen involvement in city beautification and litter control has been promoted and supported by the City of Edmonton over the years under a number of monikers. Currently, the Capital City Clean Up moniker comprehends a wide range of initiatives relating to spring cleanup, litter and graffiti management, and safe needle disposal. Programs developed by the city encourage and support citizen involvement. See http://www.edmonton.ca/programs_services/capital-city-clean-up.aspx.
23. Kingwell, *The World We Want*, 222.

Sources

Archival Sources

Much of the research for this book has been conducted in archives, including the City of Edmonton Archives, the Provincial Archives of Alberta, the Glenbow Archives, the Hudson's Bay Company Archives, and the University of Alberta Archives. In addition, research has been carried out in several special-purpose libraries including the Devonian Botanic Garden library, the Bruce Peel Special Collections Library at the University of Alberta, and the Alberta Government Library.

Records consulted at the City of Edmonton Archives include the papers of the City Commissioners, RG 11; correspondence and reports found in other record groups; and private collections including the McAfee Family fonds and the Edmonton Horticultural Society fonds. The City of Edmonton Archives maintains a large collection of "clippings files" on all aspects of civic life and on individuals who have made their mark on the city; the reports prepared by Frederick G. Todd (1907) and Morell & Nichols (1912) are located in such files. John Patrick Day's unpublished manuscript titled "Donald Ross, Raconteur and Entrepreneur" is retained in the library of the City of Edmonton Archives, as is Jane McCracken's "Walter Ramsay Florist Ltd.," a report prepared for Edmonton Parks and Recreation, Historical and Science Services, May 1976.

Records held by the Provincial Archives of Alberta include the large Gladys Reeves fonds, the Alberta Horticultural Association fonds, and several fonds related to Georges Bugnet. Although I have not cited any of the Georges Bugnet material located there, it was in the provincial archives that I saw some of the detailed and almost indecipherable notes maintained by Bugnet on the various rose crosses in his hybridizing program.

The Hudson's Bay Company Archives in Winnipeg houses a comprehensive collection of material relevant to that company's fur-trading operations in Canada. I paid one visit to these archives and subsequently made use of their microfilm loan program to read some of the Edmonton Post journals and the Edmonton District reports that are cited in this book.

The University of Alberta Bruce Peel Special Collections Library contains copies of George Bugnet's books and some of his manuscripts, as well as a small collection of correspondence and some other materials relating to his many interests and activities. This was where I found letters written to Bugnet by Horace McFarland, editor of the *American Rose Annual*.

The University of Alberta Archives houses some material relating to the work of George Harcourt, including a print version of his radio talk "Beautifying the Home Grounds," published by the Department of Extension of the University of Alberta in *The Press Bulletin* 15, no. 5 (May 9, 1930): 1–8.

The Devonian Botanic Garden has been developed primarily as a reference library for the use of staff and students in the master gardener program. It is open to researchers by appointment. In addition to reference material, it contains copies of *Kinnikinnick*, the newsletter that has been published by the Friends of the Devonian Botanic Garden since 1974; a collection of seed catalogues going back many years; and research material relevant to the development of some of the garden's feature gardens.

The Alberta Government Library, Great West Life Site, supported five government departments, including Agriculture. Although its primary purpose was to serve the needs of government employees, it was open to the public. It was there I found annual reports written by the Department of Agriculture, as well as minutes of the Horticultural Advisory Committee.

Published Sources

Al, Rodney M.B. "The EHS's Current Decade: Edmonton's Front Yards in Bloom, 1999–2008." In *A Century of Gardening in Edmonton*, edited by Maggie Easton et al., 55–57. Edmonton: Edmonton Horticultural Society, 2009.

Alderman, W.H., ed. *Development of Horticulture on the Northern Great Plains*. Institute of Agriculture, University of Minnesota, St. Paul, MN: The Great Plains Region American Society for Horticultural Science, 1962.

Anderson, Anne. *Some Native Herbal Remedies*, as told to Anne Anderson by Luke Chalifoux. Friends of the Botanic Garden of the University of Alberta, Publication Number 8. Edmonton: Department of Botany, University of Alberta, 1977.

———. *The First Métis—A New Nation*. Edmonton: UVISCO Press, 1995.

Anon. "A Peace Place." *New Trail* (Winter 1990): 24–25.

Binnema, Ted, and Gerhard J. Ens, eds. *The Hudson's Bay Company Edmonton House Journals, Correspondence, and Reports: 1806–1821*. Calgary: Historical Society of Alberta, Alberta Records Publication Board, 2012.

Brown, Jane. *The Pursuit of Paradise: A Social History of Gardens and Gardening*. London: HarperCollins, 2000.

Brown, Jennifer S.H. *Strangers in Blood: Fur Trade Company Families in Indian Country*. Vancouver: University of British Columbia Press, 1980.

Bugnet, Georges. *The Forest*. Translated by David Carpenter. Montreal: Harvest House, 1976. Originally published in the French language in Montreal by Éditions du Totem, 1935.

———. *Nipsya*. Translated by Constance Davies Woodrow. New York: Louis Carrier & Co., 1929.

———. "The Search for Total Hardiness." *American Rose Annual* (1941): 111–15.

Canada Department of Agriculture. *Canada Agriculture, The First Hundred Years*. Historical Series No. 1. Ottawa: Queen's Printer, 1967.

Carpenter, David. "Nomme de Plume." In *Writing Home: Selected Essays*. Saskatoon: Fifth House Publishers, 1994.

———. *Writing Home: Selected Essays*. Saskatoon: Fifth House Publishers, 1994.

Cashman, Tony. *Edmonton Exhibition, the first hundred years*. Edmonton: Edmonton Exhibition Association, 1979.

Chan, Marty. "Chinese Garden Reflection." *Legacy* (Summer 2009): 8–10.

"City Farmers." *The Country Guide* (June 1944): 14.

Cotterill, Patsy. "Losing Little Mountain: A Personal Anatomy of a Campaign to Save a Remnant Aspen Parkland Site." In *Coyotes Still Sing in My Valley: Conserving Biodiversity in a Northern City*, edited by Ross W. Wein, 231–48. Edmonton: Spotted Cow Press, 2006.

Day, John Patrick. "Donald Ross, Old Timer Extraordinaire." In *Edmonton: The Life of a City*, edited by Bob Hesketh and Frances Swyripa, 31–39. Edmonton: NeWest Press, 1995.

Denesiuk, Richard, and Paul Olsen, eds. *The 2008 Prairie Garden: Western Canada's Only Gardening Annual*. Canada: The Prairie Garden Committee, 2007.

Dodd, Cherry, and the Edmonton Naturalization Group. *Go Wild With Easy to Grow Prairie Wildflowers and Grasses*. Edmonton: Edmonton Natural History Club, 2004.

Dreyer, Peter. *A Gardener Touched With Genius: The Life of Luther Burbank*. New York: Coward, McCann & Geoghegan, 1975.

Easton, Maggie, et al., eds. *A Century of Gardening in Edmonton*. Edmonton: Edmonton Horticultural Society, 2009.

Evans, Clinton L. *The War on Weeds in the Prairie West: An Environmental History*. Calgary: University of Calgary Press, 2002.

Evans, David. *The City of Edmonton: History of the City Council, 1892–1977*. Edmonton: City of Edmonton, 1978.

Friesen, Gerald. *The Canadian Prairies: A History*. Toronto: University of Toronto Press, 1984.

Fry, Harold S., ed. *Development of Horticulture on the Canadian Prairies: An Historical Review*. Alberta Horticultural Association, 1986.

Gardiner, Martina. "Borden Park at 100." In *A Century of Gardening in Edmonton*, edited by Maggie Easton et al., 13–15. Edmonton: Edmonton Horticultural Society, 2009.

Grant, George M. *Ocean to Ocean: Sandford Fleming's Expedition Through Canada in 1872. Being A Diary Kept During A Journey From The Atlantic To The Pacific With the Expedition of the Engineer-in-Chief of the Canadian Pacific and Intercolonial Railways*. Toronto: James Campbell & Son; London: Sampson Low, Marston, Low, & Searle, 1873. Facsimile reprint, Coles Canadiana Collection. Toronto: Coles Publishing Company, 1970.

Harcourt, George. "Beautifying the Home Grounds." *The Press Bulletin* 15, no. 5 (May 9, 1930): 1–8. UAA, Accession No. 72-129:1–9.

Harris, Peter, and André Imbeault. "The Mysterious Parentage of 'Thérèse Bugnet'." *Roses-Canada*, no. 55 (July 2011): 3–6.

Herzog, Lawrence. *Built on Coal: A History of Beverly, Edmonton's Working Class Town*. Edmonton: Beverly Community Development Society, 2000.

Hesketh, Bob, and Frances Swyripa, eds. *Edmonton: The Life of a City*. Edmonton: NeWest Press, 1995.

Hole, Jim. *What Grows Here? Vol. 2, Locations*. St. Albert, AB: Hole's Publishing, 2004.

Hole, Lois. *I'll Never Marry A Farmer: Lois Hole on Life, Learning & Vegetable Gardening*. St. Albert, AB: Hole's Publishing, 1998.

———. *Lois Hole's Tomato Favorites*. Edmonton: Lone Pine Publishing, 1996.

Innis, Harold A. *The Fur Trade in Canada: An Introduction to Canadian Economic History*. Revised edition. Toronto: University of Toronto Press, 1970.

Johns, Walter H. *A History of the University of Alberta, 1908–1969*. Edmonton: University of Alberta Press, 1981.

Keswick, Maggie. *The Chinese Garden: History, Art and Architecture*. Revised with introduction by Alison Hardie. London: Frances Lincoln, 2003.

Kingwell, Mark. *The World We Want: Virtue, Vice, and the Good Citizen*. Toronto: Penguin Books, 2000.

Knowles, Hugh. *Woody Ornamentals for the Prairies*. Revised edition. Edmonton: University of Alberta, Faculty of Extension, 1995.

Kostek, M.A. *A Century and Ten: The History of Edmonton Public Schools*. Enlarged and updated edition. Edmonton: Edmonton Public Schools, 1992.

Lac Ste. Anne Historical Society, Archives Committee. *West of the Fifth: A History of Lac Ste. Anne Municipality*. Edmonton: The Institute of Applied Art Ltd. Educational Publishers, 1959.

Lawson, Laura J. *City Bountiful: A Century of Community Gardening in America*. Berkeley: University of California Press, 2005.

LeClaire, Nancy, and George Cardinal. *Alberta Elders' Cree Dictionary/alperta ohci kehtehayak nehiyaw otwestamâkewasinahikan*. Edmonton: University of Alberta Press and Duval House Publishing, 1998.

Leslie, W.R. "Horticulture Among the Indians of the Northern Great Plains." Chap. 2 in *Development of Horticulture on the Northern Great Plains*, edited by W.H. Alderman, 6–7. St. Paul, MN: The Great Plains Region American Society for Horticultural Science, 1962.

MacGregor, James G. *Edmonton: A History*. 2nd ed. Edmonton: Hurtig Publishers, 1975.

MacRae, Archibald Oswald. *History of the Province of Alberta*. 2 vols. Calgary: The Western Canada History Co., 1912.

Mandelbaum, David G. *The Plains Cree: An Ethnographic, Historical and Comparative Study*. Canadian Plains Studies 9. Regina: Canadian Plains Research Center, University of Regina, 1979.

McGee, Harry. "The Early Morden Rose Men." *The Rosebank Letter*, no. 16 (November 15, 1997): 5–6.

———. "Geo. Shewchuk's *Magnum Opus*." *The Rosebank Letter*, no. 29 (January 15, 2000): 8.

———. "John Alexander Wallace." *The Rosebank Letter*, no. 3 (January 15, 1996): 1–2.

———. "Octogenarian Honoured: a tribute to George Shewchuk." *The Rosebank Letter*, no. 4 (March 14, 1996): 2.

McKeague, J.A., and P.C. Stobbe. *History of Soil Survey in Canada 1914–1975*. Historical Series No. 11. Ottawa: Canada Department of Agriculture, 1978.

Meili, Dianne. *Those Who Know: Profiles of Alberta's Aboriginal Elders*. 20th Anniversary Edition. Edmonton: NeWest Press, 2012.

Merrett, Kathryn Chase. "The EHS Centennial Garden." In *A Century of Gardening in Edmonton*, 41–43. Edmonton: Edmonton Horticultural Society, 2009.

———. *A History of the Edmonton City Market, 1900–2000: Urban Values and Urban Culture*. Calgary: University of Calgary Press, 2001.

———. "Public and Private Gardens for EHS Members, 1900–2000." In *A Century of Gardening in Edmonton*, 50–51. Edmonton: Edmonton Horticultural Society, 2009.

Olsen, Paul. "Forty-five Years of Parkland Roses." In *The 2008 Prairie Garden Annual*, edited by Richard Denesiuk and Paul Olsen, 23–25. Winnipeg: The Prairie Garden Committee, 2007.

Olsen, Paul G. "Robert Simonet's Roses." *The Rosebank Letter*, no. 30 (March 15, 2000): 1–3.

———. "The Roses of Georges Bugnet." *The Rosebank Letter*, no. 34 (November 15, 2000): 1–3.

Papen, Jean. *Georges Bugnet: homme de lettres canadien*. Saint-Boniface, MB: Éditions des Plaines, 1985.

Pittao, Arnold F., and Brian J. Porter. "Parkland and Explorer Roses Cultivar Data." In *The 2008 Prairie Garden Annual*, edited by Richard Denesiuk and Paul Olsen, 30–37. Winnipeg: The Prairie Garden Committee, 2007.

Pyszczyk, Heinz W., Ross W. Wein, and Elizabeth Noble. "Aboriginal Land-Use of the Greater Edmonton Area." In *Coyotes Still Sing in My Valley: Conserving Biodiversity in a Northern City*, edited by Ross W. Wein, 21–47. Edmonton: Spotted Cow Press, 2006.

Royer, France, and Richard Dickinson. *Plants of Alberta*. Edmonton: Lone Pine Publishing, 2007.

Shewchuk, G.W. *Roses: A Gardener's Guide for the Plains and Prairies*. Revised edition. Edmonton: University of Alberta, Faculty of Extension, 1999.

Shute, Allan, and Margaret Fortier. *Riverdale: From Fraser Flats to Edmonton Oasis.* Edmonton: Tree Frog Press, 1992.

Tarlton, Fred. "The Planting and Transplanting of Lilies." *Kinnikinnick* 1, no. 3 (February 1975): 3–5.

Tingley, Ken, and Paul Bunner. *Heart of the City: A History of Cloverdale from Gallagher Flats to Village in the Park.* Edmonton: Cloverdale Community League, 2005.

Toop, Edgar. "The Marigold: Edmonton's Floral Emblem." In *A Century of Gardening in Edmonton,* edited by Maggie Easton et al., 36–37. Edmonton: Edmonton Horticultural Society, 2009.

Vick, Roger. "George Harcourt, Horticulturist." *Alberta History* 39, no. 3 (Summer 1991): 21.

von Baeyer, Edwinna. "City Beautiful Movement." In *The Canadian Encyclopedia,* edited by James H. Marsh. 2nd ed. Edmonton: Hurtig Publishers, 1988.

———. *Rhetoric and Roses: A History of Canadian Gardening 1900–1930.* Markham, ON: Fitzhenry & Whiteside, 1984.

Wein, Ross W. "Our Conservation Legacy in Edmonton." In *Coyotes Still Sing in My Valley: Conserving Biodiversity in a Northern City,* edited by Ross W. Wein, 11–20. Edmonton: Spotted Cow Press, 2006.

Wein, Ross W., ed. *Coyotes Still Sing in My Valley: Conserving Biodiversity in a Northern City.* Edmonton: Spotted Cow Press, 2006.

Wilkinson, Kathleen. *Trees and Shrubs of Alberta: A Habitat Field Guide.* Edmonton: Lone Pine Publishing, 1990.

Wilson, William H. *The City Beautiful Movement.* Baltimore: Johns Hopkins University Press, 1989.

Newspapers and Newspaper Articles

It would be difficult to overestimate the importance of local newspapers as sources of information about Edmonton's gardening history. The *Bulletin,* which began publishing as a weekly in December 1880 and evolved over time to become the daily known as the *Edmonton Bulletin,* was full of references to gardening and horticulture during the early years of the Edmonton settlement. A thorough reading of this newspaper, from its inception until the early 1900s, brought to light an intriguing picture of the role played by horticulture in the development of the city. Both the *Edmonton Bulletin* and the *Edmonton Journal* have chronicled major horticultural events in the city, events such as flower shows, garden competitions, and a host of city beautification initiatives, including in recent years the annual Communities in Bloom and Front Yards in Bloom events. In addition, both these newspapers have incorporated gardening sections and gardening columnists into their pages. Although these are not the only newspapers referred to in this book, they are the major ones. Newspaper articles, whether or not the author is named, are cited in the notes that accompany each chapter in order to make them accessible to readers.

Newspapers cited in the notes were read on microfilm collections held by the City of Edmonton Archives and the Alberta Legislature Library.

Interviews

Information supplied to the author through a combination of formal and informal interviews has been critical to the writing of this book. Interviewees not cited in the text nevertheless helped to shape the author's understanding of the many roles played by gardening during Edmonton's growth from a settlement outpost to a modern, urban conglomerate.

Formal interviews consisted of multiple meetings with one or more individuals at the same time and the preparation by the author of an interview summary that was approved and signed off by the interviewees. The preparation of these summaries almost always included e-mail correspondence with interviewees to clarify and sometimes develop points. Interview summaries prepared in this way, along with any photos and other material supplied to the author by the interviewee, will ultimately be donated to the City of Edmonton Archives. Informal interviews involved exchanging information over coffee, by e-mail, or over the telephone. In one case, the e-mail interview process lasted several months.

The names of those whose interviews are cited in the text are noted below along with particulars.

Chen, Eric, and Ruby Law. Interviews by author. February 11, 2008; March 22, 2008.
Cotterill, Patsy. Telephone conversation with author. October 10, 2012.
Craven, Joe, Kevin Hogan, and John Aiken. Interview by author. August 1, 2002.
Dorin, Walter, and Ethel Dorin. Interviews by author. November 26, 2003; February 2, 2004; March 4, 2004.
Dunn, Janine. Interviews by author. July 24, 1995; September 25, 1995; June 13, 1998.
Grenier, Antoinette. Interviews by author. August 3 and 24, 2004; November 3, 2004; February 9, 2005.
Helder, John. Interviews by author. November 18, 2002; April 9, 2003; February 27, 2004; July 24, 2012.

Hole family. Interview by author. September 24, 2004. Interview by author with staff members Bob Stadnyk, Shane Neufeld, and Bruce Keith. October 7, 2003. Interview by author with staff member Dave Grice. February 26, 2004.

Keeler, Bea. Interviews by author. June 8, 17, and 25, 2002.

Knowles, Hugh. Interviews by author. November 14, 21, and 28, 2002; January 16 and 30, 2003; February 13 and 21, 2003.

Lotzgeselle, Heiko. Interview by author. May 7, 2012.

Ng, Francis. Interview by author. April 14, 2010.

Ng, George. Interview by author. March 22, 2010; telephone conversation with author, April 6, 2010.

Paterson, Barbara. Telephone conversation with author. April 15, 2003.

Pearsell, Grant, and Angela Hobson, City of Edmonton Office of Biodiversity. Interview by author with Pearsell and Hobson. January 17, 2010. Interview by author with Pearsell. January 16, 2013.

Penstone, Susan. Interview by author. April 12, 2005.

Riske, Ken. Meeting with author at Millcreek Nursery, Edmonton. June 2012.

Roberto, Claude. Interview by author. July 2003.

Seymour, Patrick. Interviews by author. July 24, 2003; August 11, 2003; December 3, 2003.

Shewchuk, George. Interviews by author. July 4 and 24, 2002; August 9 and 27, 2002; October 24, 2002.

Shewchuk, Mary. Interviews by author. August 22, 2002; September 20, 2002; October 30, 2002.

Wigelsworth, Richard. Interviews by author. January 15 and 28, 2003.

Wong, Ging Wei. E-mail correspondence with author beginning May 2003 and extending to the present.

Wong, Kwan. Interview by author. April 2011.

Internet and Audiovisual Sources

"Annual Report of the Department of Agriculture of the Province of Alberta from the first of September, 1905, to the thirty-first of December, 1906." Edmonton: Government Printer, 1907. Accessed December 3, 2012. http://archive.org/details/annualreport1905albeuoft.

City of Edmonton. Capital City Clean Up. http://www.edmonton.ca/environmental/capital-city-clean-up.aspx.

———. Ribbon of Green Concept Plan. Accessed May 29, 2013. http://www.edmonton.ca.

———. *Urban Parks Management Plan: 2006–2016*. Accessed May 29, 2013. http://www.edmonton.ca.

Dr. Sun-Yat-Sen garden, Vancouver. Accessed September 14, 2012. http://vancouverchinesegarden.com.

Fellner, Fred. "Robert Simonet and the 'Rescued Lilies.'" *North American Lily Society Yearbook*, 1979. Manitoba Regional Lily website. Accessed March 5, 2012. http://www.manitobalilies.ca/Robert_Simonet.htm.

Government of Canada. Canada Food Board. "Report of the Canada Food Board: February 11–December 31." Ottawa, 1918. Accessed March 1, 2012. http://www.archive.org/details/cu31924013848696.

HelpMeFind. Accessed May 2, 2012. http://www.helpmefind.com/rose/l.php?l=2.2203.0.

Herzog, Lawrence. "Westmount Mall at 50." *Edmonton Real Estate Weekly, Inside Edmonton* 23, no. 35 (September 1, 2005). Accessed May 22, 2013. http://www.rewedmonton.ca/content_view_rew?CONTENT_ID=1147.

"The Honourable Lois E. Hole, C.M., 2000–2005." Government of Alberta. Accessed July 23, 2012. http://www.assembly.ab.ca/lao/library/lt-gov/hole.htm.

"Lois Elsa Hole." *CBCnews*. Accessed August 6, 2012. http://www.cbc.ca/news/obit/hole.

"Lois Hole." *Wikipedia*. Accessed August 2, 2012. http://en.wikipedia.org/wiki/Lois_Hole.

Moir, John S. "Petitot, Émile." In *Dictionary of Canadian Biography*. Vol. 14. University of Toronto/Université Laval, 2003–. Accessed November 9, 2012. http://www.biographi.ca/en/bio/petitot_emile_14E.html.

Ontario Horticultural Association. Accessed December 1, 2012. http://www.gardenontario.org.

Ragan, Philip. *He Plants for Victory*. Montreal: National Film Board, 1943. Accessed February 27, 2012. http://www.nfb.ca/film/he_plants_for_victory.

"Report of the Canada Food Board: February 11–December 31." Ottawa: Canada Food Board, 1918. Accessed March 1, 2012. http://www.archive.org/details/cu31924013848696.

Turner, George H. "Plants of the Edmonton District of the Province of Alberta." *The Canadian Field-Naturalist* 63, no. 1 (January–February 1949): 1. Accessed September 24, 2012. http://www.biodiversitylibrary.org/item/89254#page/12/mode/1up.

University of British Columbia, Indigenous Foundations. Accessed April 1, 2014. http://indigenousfoundations.arts.ubc.ca.

Index

Page numbers in italics refer to photographs.

EHS *refers to Edmonton Horticultural Society,* EAS *refers to Edmonton Agricultural Society, and* AHA *refers to Alberta Horticultural Association.*

The names of early Chinese residents are entered without inversion; other Chinese names are inverted using the Canadian style.

Aboriginal people
    early history, 40–41, 265nn10–11
    at Fort Edmonton, 42–45, 47–48, 266n16
    in *Nipsya*, 45–47
    terminology, xiv–xv
    traditional knowledge, 41, 45–47
    wild fruit, 27–28, 47–48
    wild roses, 168
Abraham, Frederick, 147

agricultural research stations. *See* Central Experimental Farm, Ottawa; research stations
agriculture and horticulture, 145
    See also gardening; market gardening
Al, Rodney, 136
Alberta, Province of
    provincial flower, 170
Alberta, Agriculture Department
    AHA collaboration, 129–30, 281n61, 282n66
    EHS collaboration, 127–29, 132
    horticultural advisory board, 131, 282n67
    research stations, 89, 131–32, 273n31
    support for amateurs, 132
    victory garden collaboration, 158
    weed control, 155

Alberta Horticultural Advisory Board, 282n67
Alberta Horticultural and Forestry Association, 293n2
Alberta Horticultural Association (AHA)
    founding of, 126, 129–31, 281n61
    mandate and programs, 99, 131, 282n66, 293n2
Alberta Horticultural Guide, 188–89
Alberta Hotel, 31–32, 264n39
Alderman, W.H., 143
Allen, H.T. (Harvey), 65, 99
Altagold corn, 80
'American Beauty' rose, 172
Anderson, Anne, 41
Anderson, John, 201
Anderson, Thomas, 50, 170
Apache Seeds, 66–67, 74
apples, 5–6, 10–11, 49, 89, 96–97, 144, 284n14
Arbor Day, 122, 124
architecture and City Beautiful movement, 105, 133
Arends, Hendrik and Cornelia, 289n36
H. Arends' Rose Gardens, 183, 289n36
asparagus, 200–01
Aspen Parkland growing area, 39
aspen trees, 38, 39, 60, 61, 256

balsam poplar, 38, 39, 60, 61
Bark Ging Wong. *See* Wong, Bark Ging
barley in HBC gardens, 42–43, 266n15
Barlow, John, 108
beans, 80, 158
beautification, city. *See* city beautification

"Beautifying the Home Grounds" (Harcourt), 59
'Beauty of Hebron' potatoes, 25–26, 28, 262n12
Beaverlodge research station, 49, 89, 172, 189, 273n31
Beech, Henry, 222
beekeeping, 4, 5, 168, 170, 213, 260n9
beets, 129, 158, 198, 222, 286n46
berry fruits
    abundance of wild fruit, 10, 27–28, 39–40, 45, 47, 51
    native plants, 39, 45, 47, 51
    wild *vs.* cultivated distinction, 26–28, 48, 50–51
    See also specific fruit
Binnema, Ted, 42
birch, 38–39, 58–59, 122–24
Blake, G.A., 26
Blakey, William G., 280n45
Blatchford, Kenneth, 122
blueberries, 27, 39, 45, 51
Blyth, Mrs., 173
bok choy, 219–20, 222–23, 230
Borden Park (East End Park), 56, 57
Brandon research station, 87, 186, 273n31
breeders, early plant, 81
    See also Bugnet, Georges; Simonet, Robert
Brett, Robert G., 124, 142, 283n8
'Bride' rose, 70
'Bridesmaid' rose, 70
Brintnell area (Little Mountain), 60–61, 260n9, 269n65, 269n67

Brooks research station, 89, 131–32, 273n31
Brown, Ernest, 8, 35, 43, 118, 122, 125
Brown, Jane, 37–38
Buchanan, George, 114, 246, 294n5
Bugnet, Georges, *83*, *85*
   about, 11–12, 82–84
   Aboriginal people and, 45
   family life, 83–84, 86
   The Forest, 11–12
   Nipsya, 45–47, 89, 266n26
   plant breeder, 12, 87–89, 185–87
   publications and awards, 86
   roses, 80–81, 88–89, 185–87
   "The Search for Total Hardiness", 90
   seed and plant sources, 12, 87–88
   See also 'Thérèse Bugnet' rose
Bugnet, Georges, views
   climate and breeding, 88–89
   hardiness of Alberta plants, 87, 89
   nature and the divine, 12, 86–87, 90, 186, 261n27, 273n44
   'Thérèse Bugnet' rose, 81–82, 89–90
   war, 90
   wealth, 82, 88
Bugnet, Julia, 82–84, *83*
Bugnet, Maurice and Charles, *83*
Bugnet, Thérèse, 86, 261n28
Bugnet Scots pine, 88, 274n55
Bulletin. See *Edmonton Bulletin*
Bulyea, George, 103
Bunner, Paul, 292n11
Burbank, Luther, 25, 32–33, 87, 262n13
Bury, Ambrose, 112, 113
Buswell, Arthur, 149
Butler, Patrick, 235, 293n18

cabbages
   Chinese gardeners, 198, 204, 215, 218–19, 222–24
   HBC gardens, 24, 42, 266n15
   seed supplies, 74
   vacant lots program, 129
   victory gardens, 158
Calgary
   market garden shipments to, 31
   vacant lot gardening, 142
   weed control, 155
Canada
   Chinese immigration, 198, 206–08, *207*, 225
Canada, Agriculture Department
   Canada Food Board, 144–48
   research stations, 49, 89, 273n31
   support for Edmonton research, 49–50
   vacant lot gardening, 144–48
   wartime farm workers, 145
   See also Central Experimental Farm
Canada Post, 135–36, 254
Capitol Theatre, 12, 176
Capitol Theatre rose shows, 13, 78, 177–79, 181–82, *182*, 261n32
caragana, 89
Cardy, William J. (Bill)
   AHA role, 126, 129–30, 281n61, 282n66
   Cardy Shield award, 281n62
   as citizen gardener, 246
   city beautification activist, 104, 117–18, 276n7

EHS role, 126–30, 159, 246
life of, 126–27, 130, 280n55
personal traits, 118, 126–27, 128
vacant lots program, 110
carnations, 69–70, 171
Carpenter, David, 65, 81–82, 86, 99
carrots, 80, 129, 158, 204, 286n46
Carters Seeds, England, 8, 10, 80
Cashman, Helen Gorman, 200–02
Cashman, Tony, 292n11
catalpa, 50
cauliflower, 19, 218–19
Cavanagh, Terry, 199
cedar trees, 58
cenotaph, 113, 278n27
Centennial Garden in Henrietta Muir
  Edwards Park, 114, 116–17, *117*,
  135, 194, 279n34
Centennial Rose Garden in Coronation
  Park, 114
Central Experimental Farm, Ottawa
  about, 6, 26, 49–50
  breeding programs, 26
  strawberries, 7, 25, 26, 50
  wild *vs.* cultivated distinction,
    26–28, 48
  See also research stations
A Century of Gardening in Edmonton (EHS),
  136
Chalifoux, Luke, 41
Chalmers, Ron, 60–61, 269n65
Chambers, Allan, 60
Chan, Marty, 197, 200, 234
Chen, Eric, 232–34
Chen's Market Garden, 233
cherries, 25, 160

Cheung Gee, 217
Chew Lung, 210
children
  Chinese garden work, 214–18,
    223–25
  HBC garden work, 42–43, 266n16
  Reeves as children's activist, 120,
    244
  in shows, 244
  See also schools
Chin Lock, 201
Chinatown, 199, 229, 291n3
Chinese community
  Canadian and Chinese names, 201,
    291n5
  discrimination against, xi, 198, 225
  immigration, 198, 206–08, *207*, 225
  language learning, 206, *207*, 225
  traditional culture, 225–27, 229–30,
    235–38, *236–37*
  See also Wong, Bark Ging; *and other
    Wong family members*
Chinese Garden, Louise McKinney
  Park, 198–200, 234–36, *236–37*,
  291nn2–3, 293nn18–19
The Chinese Garden (Keswick), 235
Chinese market gardening
  about, xiii, 206
  discrimination against, xi, 206
  early history, 200–02
  locations of gardens, 198, *198*,
    200–05, *203*, *205*, 215–17
  produce and markets, 198, 202,
    204–06
  See also Chen, Eric; Wong, Bark
    Ging, market gardening

Chinese vegetables
   organic gardening, 233–34
   Wong's gardens, 15, 219–20, 222–23, 230–31
chokecherries, 25, 39, 41, 45, 96, 97, 160
Chow, Bob and Flora, 216
choy sum, 231
chrysanthemums, 70
citizen gardeners
   caring and citizenship, 251, 258
   city beautification, 136, 244
   Front Yards in Bloom, 136
   leadership qualities, 244–45
   for social change, 243–44, 246–47, 251
   teamwork, 246
   values of, 241–43, 251, 258
   See also Helder, John; Hole, Lois; Reeves, Gladys
city beautification
   city practical *vs.* city beautiful, 105
   commemorations, 124, 280n45
   Communities in Bloom, 107, 132–37, *134*, 254, 282nn70–71, 295n22
   community garden movement and, 163–65
   Front Yards in Bloom, *134*, 135–36, 254
   physical and moral beauty, 105–06, 136–37
   social action *vs.* lifestyle option, 109
   See also cleanup campaigns; Edmonton, City of, collaboration with EHS; Edmonton Horticultural Society, city beautification; vacant lots and community gardening
City Beautiful movement
   about, 103–04, 136–37
   architectural focus, 105, 133
   collaboration of EHS and city, 106–07, 109
   comparison with Communities in Bloom, 133
   decline of, 109
   international scope, 103–05
   Keomi Club support for, 120, 294n5
   leadership for, 118–20, 125–26
   Reeves on, 101
The City Beautiful Movement (Wilson), 103–05
City Bountiful (Lawson), 141
City Farm, 287n55
City of Edmonton. *See* Edmonton, City of
Civic Block offices for EHS, 110–11, 113, 277nn18–19
Clarke, Joseph, 151
'Claude Bugnet' plum tree, 89
cleanup campaigns
   citizen gardners, 241–42
   collaboration of EHS and city, 109, 111
   history of, 111
   publicity for, 293n1
   recent scope of, 295n22
   Reeves's role, 122–23, 242, 246
   tree planting, 122–23
   vacant lots program, 147–48
climate
   boosterism and, 39–40

local *vs.* research stations, 87
optimism of gardeners, xii
temperature variations, 49
Clover Bar bridge area, 215–16
commemorative benches, 124
Communities in Bloom, 107, 132–37, 134, 254, 282nn70–71
Community Garden Network, 162–64
community gardening. *See* vacant lots and community gardening
Community Gardens, 287n55
community leagues
  floral colour scheme for parks, 245–46
  Front Yards in Bloom, 135
  tree planting and, 111, 119, 122, 293n2
Connaught Drive beautification, 113
corn, 80, 158, 204
Coronation Park, 114
Cotterill, Patsy, 60–62, 269n63, 269n67
cranberries, 27, 39, 51, 53, 266n26
Cree people
  early history, 41, 265nn10–11
  in *Nipsya*, 45–47
  See also Aboriginal people
Cromdale Community League, 245
cucumbers, 28, 249
culture and nature
  about, xii, 37–38
  Aboriginal *vs.* non-Aboriginal views, 45–48
  balance of protection and transformation, 56–62
  integrated gardens, 51–53
  integration of concepts, 255

property rights and, 43–45
recent trends, 38, 255–58
research stations and views of, 49–50
soil quality, 39
wild *vs.* cultivated distinction, 48
See also gardening
culture and nature, history
  fur trade era, 42–48
  in 1880s and 1890s, 48–52
  in early 1900s, 53–58, 102
  parks policies, 53–54
currants, 5, 6, 27, 39, 45, 50, 51, 52, 154

dahlias, 127
Day, John Patrick, 22, 33
de Cazes, Count, 29
delphinium, 245
The Development of Horticulture (Alderman), 143
Devonian Botanic Garden
  about, 131
  classes, 189
  Kurimoto Japanese Garden, 199, 291n1
  Native Peoples Garden, 41
  reference library, 298
  rose garden, 190
  Simonet's lilies, 93
Dickinson, Richard, 38
displays. *See* shows, exhibitions, and displays
Dixon, Charles, 244
Dolgoy, Max and Rachael, 226

Dominion Experimental Farm. *See*
    Central Experimental Farm,
    Ottawa
Dorin, Ethel, 160
Dorin, Walter, 77, 80, 160, 270n3,
    272n24
Dreamland Theatre, 225
Duchesne, Catherine, 163
Dunn, Janine, 91–93, 98, 271n4
Durocher, Virginie, 248

'Earl Haig' rose, 179
'Early Rose' potatoes, 25, 26, 28
East End Park (Borden Park), 56, 57
Edmonton area, early history
    presettlement period, 38–42
    fur trade era, 1–2, 42–48
    land surveys, 20–21
    See also Aboriginal people
Edmonton, City of
    floral colour scheme, 245–46
    Harbin twin city, 238
    marigold as official flower, 194–95
    nursery, 257
    Strathcona amalgamation, 111
    See also cleanup campaigns; parks
Edmonton, City of, collaboration
    with EHS
    about, 106–07, 109
    Centennial Garden, 116–17, 117
    city beautification mandate, 110–14,
        116
    Civic Block offices, 107, 109, 110–11,
        113, 277nn18–19
    floral colour scheme for parks,
        245–46

funding to city from EHS, 113–14,
    116
funding to EHS from city, 109,
    111–13
Muttart garden, 114, 116
Partners-in-Parks program, 116–17,
    162–63, 253–54
proposal to merge EHS with City,
    109–10
relief during Depression, 111, 148–
    51, 284n23, 285nn25–26
See also cleanup campaigns;
    Communities in Bloom; Helder,
    John; vacant lots program
Edmonton Agricultural Society (EAS),
    2–3, 28–30, 49
    See also Edmonton Horticultural
    Society
Edmonton and Area Land Trust, 61,
    269n68
Edmonton Bulletin
    gardening boosterism, 2–7, 10, 19,
        39, 157, 178–79, 288n5
    named varieties, 25–26
    research coverage, 6–7
    soil quality, 39
    trial garden, 5–6
    wild *vs.* cultivated distinction,
        26–28
Edmonton Chinese Garden Society,
    199–200, 234–35, 291n2
Edmonton City Market. *See* Market
    Square
Edmonton Exhibition Association, 129,
    281n60

Edmonton Horticultural and Vacant
    Lots Garden Association, 108
Edmonton Horticultural Society
    budgets and finances, 111–13, 116,
        119, 132, 156
    bylaws, 108
    founding of, 34, 102–03, 106–07
    honorary members, 103
    leadership quality, 107, 276n7
    mandate, 106–09
    name changes, 108–09, 143, 259n1,
        276n3
    publicity, 127–28
    speaker program, 132
    See also Edmonton, City of,
        collaboration with EHS; shows,
        exhibitions, and displays
Edmonton Horticultural Society, city
    beautification
    about, 101–04, 107
    Centennial Garden, 116–17, 117, 135,
        194, 279n34
    Front Yards in Bloom, 134, 135–36,
        254
    leadership, 104, 106–07, 117–18
    mandate for, 102–03, 105–09, 112,
        132–33, 276n7
    model garden, 113
    Muttart Conservatory garden, 114,
        116, 117, 135, 194
    plant donations, 113–14
    provincial funding for, 132
    See also City Beautiful movement;
        cleanup campaigns;
        Communities in Bloom

Edmonton Horticultural Society,
    historical overview
    amalgamation with Strathcona
        society, 108, 111, 276n3
    first female president, 118
    founding of, 34, 102–03, 106–07
    1909 to 1930s, 102–04, 106–08
    1930s to 1950s, 103, 127–31
    1950s to 1990s, 131
    1990s to 2010s, 132–37
    See also Cardy, William J. (Bill);
        Reeves, Gladys
Edmonton Hotel, 22–24, 23, 27, 260n16
Edmonton House, terminology, 266n15
    See also Fort Edmonton
Edmonton Journal
    gardening boosterism, 10
Edmonton Market Square. See Market
    Square
Edmonton Native Plant Group, 61–62,
    135, 256, 270n69
Edmonton Nature Club, 60–61
Edmonton Tree Planting Committee,
    59, 123
    about, 122, 293n2
    citizen gardeners, 244, 246
    cleanup campaigns, 122–23
    collaboration of EHS and city, 109,
        111–12, 119, 122
    commemorations, 124
    community leagues, 111, 119, 122
    funding for, 112, 124–25
    native species, 58–59, 123–25
    nursery for, 58, 122, 125
    Reeves's role, 58–59, 120–25, 121
    statistics on plantings, 125

85th Avenue, 59, 124
Elizabeth Avenue, 67–69
elm trees, 58
Enjoy Centre, 66–67
Ens, Gerhard J., 42
environmental stewardship
    Communities in Bloom program, 132–33, 254–55
    community garden movement, 161–64
    insect and animal pests, 174, 221–22
    natural area parks, 56, 60–61, 269n68
    naturalist program, 255–56
    Naturescapes program, 255, 257
    organic gardening, 232–34
    See also cleanup campaigns; Little Mountain; native plants; weed control
ethnicity and race. *See* Aboriginal people; immigrant gardeners; *and entries beginning with* Chinese
Evans, Clinton, 37–38, 154–55
exhibitions. *See* shows, exhibitions, and displays
Explorer roses, 96, 188–89, 191, 194

Fellner, Fred, 93–94, 98
Ferguson's Scotch Roses, 174
Finlay, Cam and Joy, 60
First Nations people, terminology, xiv–xv
    See also Aboriginal people
First Presbyterian Church, 126
First World War
    cenotaph, 113, 278n27

food shortages, 145
    vacant lot gardening, 141–42, 144–46, 161
    wartime farm workers, 145
Fleming, Sandford, 43–44
Flora of Alberta (Moss), 38
Florists Transworld Delivery (FTD), 70, 73
flowers. *See* gardening; market gardening; Ramsay, Walter
food security, 145, 148–49, 164
    See also vacant lots and community gardening
Fook Kwan Wong. *See* Wong, Fook Kwan
The Forest (Bugnet), 11–12
Fort Edmonton, 44
    Aboriginal people at, 42–45
    farm and garden, 22, 24, 42–44, 262n8, 266n15
    land surveys, 20–21
    languages at, 266n16
    shows and exhibitions, 2–4, 28
    terminology, 266n15
    See also Hudson's Bay Company
Fort Edmonton Park, 260n9, 270n1
Fort Vermilion research station, 49, 273n31
Fortier, Margaret, 202–04
Front Yards in Bloom, *134*, 135–36, 254
fruit trees and bushes
    Bugnet's research, 89
    early history of research, 6
    Harcourt's research, 144
    history of, 51
    integrated gardens, 53

newspaper promotion, 5–6
Simonet's research, 96–97
sources, 49
See also apples; berry fruits; pear trees; plum trees
fuschias, 15, 33, 171

gai choy, 219–20, 223, 230–31
gai lan, 219, 222, 231
gardening
    beauty and human behaviour, 105–06, 242
    Bugnet on nature and the divine, 86–87, 90, 186, 273n44
    caring as soul of, 247
    desire for beauty, 14–15, 136–37
    desire for challenge, 11, 13–14
    desire for the divine, 12, 186, 261n27
    desire for the familiar, 11, 15, 168, 186
    early bias against western ways, 2, 4, 15–16
    early boosterism for, 2–7, 10–11, 19, 39, 157, 178–79, 288n5
    "from Nature's garden," 41, 47, 51
    as healthy individual activity, 109
    integrated gardens, 51–53, 255
    meditative and restorative activity, 186
    as metaphor for life, 16, 248–49
    naturescapes, terminology, 255
    optimism of gardeners, xii, 16
    as political act, 104, 106, 109, 141, 160, 164–65
    private as public, 136
    purchased *vs.* homegrown preferences, 159–60
    social action *vs.* lifestyle option, 109
    "Why grow here?," 16
    See also citizen gardeners; culture and nature
Gardiner, Martina, 56
gathering. *See* hunting and gathering
Gee Gut, 201–02, 217
'George Dickson' rose, 177
George G. Ball Co., 93
George Hop, 203
Georges Bugnet (Papen), 270n2
Ging Wayne Wong. *See* Wong, Ging Wayne
Ging Wei Wong. *See* Wong, Ging Wei
gladioli, 14, 96, 97, 127, 160, 167, 171–72
'Gloire de Dijon' rose, 178
gooseberries, 4–5, 6, 15, 27, 45, 50, 51, 154, 220
Government House, 211–12
government roles. *See* Alberta, Agriculture Department; Canada, Agriculture Department; Edmonton, City of
grains in HBC gardens, 42–43, 266n15
Grand Trunk Park, 215, 293n14
Grant, George, 43–44
Great Depression
    EHS activities, 127–28
    food shortages, 148–49
    impact on Ramsay, 73
    relief collaboration of EHS and city, 111, 148–51, 284n23, 285nn25–26
    vacant lot gardening, 148
greenhouses and hotbeds

family businesses, 66–67
first greenhouses, 50
horticultural standards, 66–67
hotbeds in Chinese gardening, 215, 218–19, 221, 231
hotbeds in Fort Edmonton, 42
Marriaggi's greenhouse, 31–32, 264n39
Ramsay's greenhouses, 34, 66–70, 69, 73, 270n1
Ross's greenhouses, 31–34
water supplies, 32
See also market gardening
Greenland Garden Centre, 66–67
Grenier, Antoinette, 154
Groat Ravine market gardening, 202, 203, 211
Groat Ravine Park, 54–55, 55
Growing Roses in the Prairie Provinces (Shewchuk), 287n1
Gum Lang, 209

'Hansa' rose, 178
Hansen, N.E., 87–88, 274n53
Harcourt, George, 142
   apple breeding, 144, 284n14
   career, 35, 59, 143
   as citizen gardener, 246
   city beautification activist, 59, 104, 246, 276n7
   native plants, 59
   personal traits, 143–44
   publications, 59
   vacant lots program, 59, 143–44, 146, 277n15
   viability testing by Pike, 77, 272n23

Hardie, Alison, 235
Harry Yee, 215–16
hawthorn, 89
HBC. *See* Hudson's Bay Company
He Plants for Victory (film), 139–40, 157
Heart of the City (Tingley and Bunner), 292n11
Helder, John
   about, 251–52
   as citizen gardener, 251–58
   city beautification, 254–55
   Communities in Bloom, 134–35, 282n71
   community garden movement, 162–63, 253–54
   Naturescapes program, 255, 257
   principal of horticulture, 114, 116, 134–35, 162, 251–56, 282n71
   promotion of partnerships, 116, 162–63, 252–53
hemp, 14
Henderson, Peggy, 12, 170–71, 178
Henderson, Thomas, 4–5, 170–71, 260n9
Henderson's Directories, 203
Henrietta Muir Edwards Park, 116, 279n34
herbicides, 256–57
Hilliard, Fred, 96
Hodgson, Ernest, 181–82
Hole, Bill and Jim, 16, 248
Hole, Lois
   about, 16, 246–47
   as citizen gardener, 246–51
   market gardening, 247–49
   public service, 250–51

publications, 248–49
Hole, Lois, views
    caring as soul of gardening, 247, 251
    gardening as metaphor, 16, 248–49
    studying and doing, 249–50
    "Why grow here?," 16
Hole, Ted, 246–48
Hole's Greenhouses and Gardens, 247–49
Holland, H.A., 174
hollyhocks, 96
Holme, Isobel, 176
Home, John, 179
honey and beekeeping, 4, 5, 168, 170, 213, 260n9
honeysuckle, 89
Hop Woo, 203–05, 209, 210
hops, 39
Horticultural Advisory Board, 131
horticulture and agriculture, 145
    See also gardening; market gardening
hotbeds. See greenhouses and hotbeds
Hotel Macdonald, 119
Housley, Thomas, 150
Hudson's Bay Company
    farms and gardens, 22, 42, 266n15
    fur trade, 42–45
    in *Nipsya*, 46–47
    See also Fort Edmonton
hybridization
    Burbank's influence, 25, 262n13
    See also roses

I'll Never Marry a Farmer (L. Hole), 248–49, 294n11

immigrant gardeners
    as cultivators, not gatherers, 27
    desire for the beauty from home, 168–69, 185–86
    desire for the familiar, 11–12, 15
    Laotians, 232–34
    roses, 12
    See also entries beginning with Chinese
Indian, terminology, xiv–xv
    See also Aboriginal people
Indian Head research station, 87, 273n31
invasive species, 155–56, 256–57
    See also weed control
Inventory of Environmentally Sensitive and Significant Natural Areas, 60, 269n63

Jalbert, Brad, 191
Japanese Garden. *See* Kurimoto Japanese Garden
Jasper Avenue, 76
Joe Wong. *See* Wong, Joe
John Janzen Nature Centre garden, 257
Juchli, Peter, 91
Jung Suey, 201

Keeler, Bea and Roy, 97, 115, 126, 160–61
T.R. Keith & Co., 48
Keomi Club, 120, 294n5
Keswick, Maggie, 235
Kingwell, Mark, 251, 258
Kinnaird, George J., 267n45
Kinnaird Ravine (Rat Creek Ravine), 53, 58, 245, 267n45, 268n57
Kinsmen Park, 114, 201–02, 292n7

Knott, Daniel, 110, 114, 284n23
Knowles, Hugh, 97–98
Knowles, J., 4–5, 50
Kurimoto Japanese Garden, 199, 291n1
Kwan Wong. *See* Wong, Fook Kwan

Lacombe research station, 49, 99, 273n31
'Ladoga' pine (*Pinus sylvestris*), 88, 274n55
Laotians, 232–34
Law, Ruby, 233–34
Lawson, Laura J., 141
Lee, Hong, 200, 202, 203
LeFeuvre, John, trophy, *173*, 174
legion halls, 280n45
Leslie, W.R., 41
Lessard, P.E., 122
Lethbridge research station, 49, 273n31
lettuce, 33, 69, 204
lilies, xv, 33, 93–94, *95*, 98, 170, 264n51
Lilium philadelphicum, 94, 264n51
Lilly Mee Wong. *See* Wong, Lilly Mee
'Lily Simonet' lily, 94, *95*
Little, J.B., 203
Little Green Thumbs, 287n55
Little Italy Business Association, 291n3
Little Mountain, 60–61, 269n65, 269n67
Living Prairie Museum, Winnipeg, 60
locust trees, 50
Lotzgeselle, Heiko and Carol, 190–93, *192*
'Louise Bugnet' rose, 187
Louise McKinney Park, 193–94, 199–200

Louise McKinney Park, Chinese Garden, 199–200, 234–36, *236–37*, 293nn18–19
low prairie rose (*Rosa arkansana*), 186, 287n3
Lychee Gardens, 229

MacDonald Drive, 280n45
Macdonald Hotel, 119
Mackenzie and Mann Park (Oliver Square), 194
Macoun, John, 48–49
MacRae, Archibald, 22, 271n5
Mah Foo, 215–16
mangold wurzel, 25, 28, 263n26
maple trees, 49, 51, 58
'Marie Bugnet' rose, 89, 189
marigolds, 15, 50, 194–95
market gardening
  early history, 5, 31–34
  Hole's Greenhouses, 247–49
  organic gardening, 233–34
  racial discrimination, xi, 206
  Ross's business, 30–34, 171
  Simonet's, 91–92
  See also Chinese market gardening; greenhouses and hotbeds
Market Square
  beautification of, 110, 113
  market gardeners, xi–xii, 34, 91–92, 206
  model garden, 111, 113
  tree sales, 122, 125
Marriaggi, Frank, 31–32, 264n39
Marshall, Henry Heard, 186, 188

McAfee, James and Hilda, 114, 157–58, 180, 181–83, 182, 261n32
McCalla, P.D. (Pete), 131, 170, 282n66
McCracken, Jane, 73, 270n1
McDougall, George, 20–21
McDougall, John A., 103
McFarland, J. Horace, 88–89, 187
McGee, Harry, 168
McKee, H.F. (Harold), 148–51, 284n23, 285nn25–26
McKenzie Seeds, 73–74, 77, 270n3
McKinney Park. *See* Louise McKinney Park
McNeil, Ted, 231
meadow rose (*Rosa blanda*), 186
Memorial Hall beautification, 124, 280n45
Métis people
  in *Nipsya*, 45–47
  terminology, xiv–xv
  See also Aboriginal people
Mill Creek Thistle Patrol, 256–57
Mill Woods housing co-op garden, 162–63, 253
Millcreek Nursery, 96
'Millstream' apple, 96
moose berries, 266n26
Morden research station, 41, 89, 186, 188, 275n67
Morell & Nichols, 53–54, 105
Moss, Ezra H., 38
moss roses, 12
mountain ash, 53, 58, 97
Muttart, Merrill and Gladys, 114
Muttart Conservatory, 252, 292n11

Muttart Conservatory garden, 114, 116, 117, 135, 194

Native Peoples Garden at Devonian Gardens, 41
  See also Aboriginal people
native plants
  economical benefits, 51
  Harcourt's views on, 59
  integrated gardens, 51–53, 255
  natural area parks, 56, 60–61, 269n68
  naturalization strategy, 256–57
  presettlement period, 38–42
  public gardens, 41, 257
  Reeves's views on, 124
  Tree Planting Committee's use of, 58–59, 123–25
  wild rose as provincial flower, 170
  wild *vs.* cultivated distinction, 26–28, 48, 50–51, 168
  See also culture and nature; Edmonton Native Plant Group; environmental stewardship
native plants, specific
  lilies, 264n51
  shrubs, 39
  trees, 38–39, 49–50, 58–59, 123–25, 256
  wild strawberries, 27–28, 47–48, 51
  See also roses, wild
nature and culture. *See* culture and nature
Naturescapes program, 255, 257
'New Canadian Queen' rose, 70
New World Restaurant, 229

Ng, Francis, 200, 235–36, 238, 293nn18–19
Ng, George, 199–200, 235, 291n2
Nipsya (Bugnet), 45–47, 89, 266n26
North Saskatchewan River Valley Area Redevelopment Plan, 54
Norway spruce, 4, 50
noxious weeds, 155–56
    See also weed control

oak trees, 15, 49, 58
O'Farrell, M.J., 146, 148–49
Old Strathcona Farmers' Market, 234
Oliver, Frank, 2–5, 28, 48
    See also *Edmonton Bulletin*
Oliver Tree Nursery, 48, 88
Olmsted, Frederick Law, 104–05
Olsen, Paul, 95–96, 187–88
100A Street, 76
Onion Park (Grand Trunk Park), 293n14
onions, 19, 25, 26, 198, 293n14
Osler, Andrew, 28, 40, 265n9

Palatine Roses, Ontario, 191
Papen, Jean, 82, 84, 86–87, 270n2
Paramount Theatre, 176, 194, 261n32
Parkland roses, 188–89, 191, 194
parks
    balance of protection and transformation, 56–62
    City Beautiful movement, 103–06
    city departments for, 54, 61
    early history, 35, 52–56, 55
    facilities in, 55–56
    floral colour scheme, 245–46
    integrated gardens, 51–53, 255

natural area parks, 56, 60–61, 269n68
Olmsted's influence, 104–05
Partners-in-Parks program, 116–17, 162–63, 253–54, 256–57
See also Centennial Garden; Louise McKinney Park; river valley parks system
parsley, 223
parsnips, 129, 158
Partners-in-Parks program, 116–17, 162–63, 253–54, 256–57
Paterson, Barbara, 176
Peace Garden, 194
'Peace' rose, 181–83, *182*, 289n29
pear trees, 49, 89, 97
Pearsell, Grant, 61, 269n68
peas, 24, 50, 80
Peas on Earth, 233–34
Pember, J.E., 147–48
Penstone, Susan, 163
peonies, 280n55
Persian rose (*Rosa foetida persiana*), 171, 178
Personal Community Support Organization, 287n55
pests, insect and animal, 174, 221–22
Petitot, Father Émile, 83–84
Petrie, Douglas, 52
petunias, 92–93, 98, 114, 187
Phair, Michael, 136
Phetsavanh family, 232–34
Pike, Alfred
    about, 80–81
    city beautification activist, 104, 276n7

EHS activities, 78, 80, 272n25
life of, 74–76, 75, 185, 272n20
publications, 12–13, 78, 80
"Rose City" campaign, xiii, 12, 78, 80
roses, 13, 78, 80, 172, 177–78, 185
seed and plant sources, 13, 76, 80
viability testing, 77, 80
Pike, Alfred, Jr., 77, 270n3, 272n24
Pike Seeds
founding of, 35, 66, 76
reorganization, 77, 272n24
retail store, 76, 76–77, 79
sale to McKenzie Seeds, 73–74, 77, 270n3
sales manager, 160
Pinus sylvestris ('Ladoga' pine), 88, 274n55
plant breeders, early
about, 81
See also Bugnet, Georges; Simonet, Robert
Plants of Alberta (Royer and Dickinson), 38
plum trees, 89
poplars, 38, 39, 53, 61
poppies, 245
potatoes
HBC gardens, 24, 42–44, 266n15
named varieties, 25–26, 28, 262n12
relief programs, 127
Ross's potatoes, 19, 25–26
seed supplies, 74
vacant lots program, 129
victory gardens, 158, 286n46
A.E. Potter & Co., 70, 76

Power Plant, 128
prickly rose (*Rosa acicularis*), 89, 168–69, 169, 186–87, 287n3
The Pursuit of Paradise (Brown), 37–38

Queen Elizabeth Park, 56

race and ethnicity. *See* Aboriginal people; immigrant gardeners; *and entries beginning with* Chinese
radishes, 80, 204, 218
Ragan, Philip, 140
Railway Barns, 127
Ramsay, Donald, 73, 271n18
Ramsay, Walter
about, 66–67, 73
EHS role, 34, 102
family life, 67–69, 73
greenhouses, 34, 66–70, 69, 73, 102, 270n1
international deliveries, 70, 73
public service, 70, 73
retail florist, 34, 66, 70–73, 71–72, 171
roses, 70, 171
raspberries, 6, 27, 39, 51, 96, 220
Rat Creek Ravine (Kinnaird Ravine), 53, 58, 245, 267n45, 268n57
Reeves, Gladys
about, 57, 118, 242–43, 276n7
as children's activist, 120, 244
as citizen gardener, 242–46
city beautification activist, 57–58, 67, 104, 118–19, 125–26, 244–45, 246, 276n7, 293n2
cleanup campaigns, 119, 122–23, 242, 246

EHS role, 57, 118–19, 125–26, 246
floral colour scheme for parks, 245–46
life of, 7–8, 118, 126
personal traits, 118, 119, 125–26
photographer, 118, 123, 276n7
Rat Creek bridge, 268n57
speaker, 57, 118–20
tree planting committee, 58–59, 120–25, 121, 245
Reeves, Gladys, views
balance of protection and transformation, 57–58
beautification, 101–02
civic passivity, 243
her contributions, 125
her father, 10
native species, 124
Ross, 8
tree planting, 120, 124
Reeves, Hugh, 244
Reeves, William Paris (Gladys's father), 1, 7–10, 9, 118, 243, 294n3
Reinert, Ed, 218
Renfrew Fruit and Floral Co., 6, 26, 48, 170
"Report of the Parks Department for 1912" (Von Aueberg), 267n48
"Report on Parks and Boulevards" (Todd), 53–54
research stations
about, 49–50, 81
climate differences between local aras and, 87
federal, 89, 273n31
provincial, 89, 131–32, 273n31

roses, 185, 186
See also Central Experimental Farm, Ottawa
rhubarb, 96
Ribbon of Green Concept, 54
Riske, Ken, 96
river valley parks system
balance of protection and transformation, 57–60
early history, 53–54, 104–05
gardens, 116
Olmsted's influence, 104–05
Riverdale, Chinese market gardeners, 202–05, 205, 210–11
Riverdale (Shute and Fortier), 202–04
Riverside Park (Queen Elizabeth Park), 56
Riverview Pavilion, 119
'Robert Simonet' lily, 94
Roberto, Claude, 185–86
Robertson Seeds, 66–67, 74
Rosa acicularis (prickly rose), 89, 168–69, 169, 186–87, 287n3
Rosa arkansana (low prairie rose), 186, 287n3
Rosa blanda (meadow rose), 186
Rosa foetida persiana (Persian rose), 171, 178
Rosa kordesii, 96, 188
Rosa woodsii (Wood's rose), 89, 186, 287n3
Rose Gardening on the Prairies (Shewchuk), 287n1
roses
about, 13–15
breeding for hardiness, 170

Bugnet on roses, 89
campaign for "Rose City," xiii, 12,
   78, 80, 172, 176–79, 181, 185,
   194–95, 289n36
classes on, 172, 189–91
cultivated *vs.* native, 168–69
Explorer series, 96, 188–89, 191, 194
hybrid roses, 13, 88–89, 170, 172,
   174, 185–90
immigrants and cultivated roses,
   12, 168–69
natural pesticides, 174
Parkland series, 188–89, 191, 194
in private gardens, 13, 188, 190–93,
   192
provincial flower, 168, 170
in public gardens, *13*, 114, 116, 188,
   193–95
publications on, 12–13, 78, 168,
   188–89, 287n1
sources for, 13, 48, 80, 170–71, 174,
   191, 193
winter protection, 179, 191, 193
See also Bugnet, Georges; Shewchuk,
   George; 'Thérèse Bugnet' rose
roses, shows
Capitol Theatre shows, 13, 78,
   177–79, 181–82, *182*, 261n32
EHS prize lists and awards, 78, 80
history of, 172–85
newspaper promotion, 178
Paramount Theatre shows, 176, 194,
   261n32
prizes and trophies, *173*, 174, 181–82,
   182, 261n32, 272n26
roses, wild

about, 168–70
abundance of, 89, 168–70, 288n5
breeding for hardiness, 170
Bugnet's rose breeding, 89–90,
   186–87
honey, 168, 170
provincial flower, 170
species, 89, 168–70, *169*, 287n3
Roses (Shewchuk), 13, 168, 287n1
Ross, Donald, *20–21*, 34
about, xii, 6–7, 19–20, 33–34
Burbank's influence on, 32–33
early life, 21–22
EAS role, 28–30
experimental farm research, 6–7,
   25, 50
greenhouses, 30–33, 171
hotel proprietor, 22–24, *23*, 27,
   260n16
market gardening, 30–34, 171
pork packing, 29
reputation, 20, 30–31, 35
river flats, *21*, 21–22, 34
seed and plant sources, 6–7, 24–25,
   50
shows and displays, 24, 25, 28–30,
   34, 35, 50
Ross, Donald, plants
flowers and houseplants, 33, 171
named varieties, 25–27
strawberries, 7–8, 25, 26–27, 50
vegetables, 19, 25–26
Rossdale, early history, *21*, 21–22, 34
Royal Glenora Club, 131, 292n11
Royer, France, 38
rugosa roses, 172, 178, 183, 187

Russian thistle, 155

St. Albert Mission, 51–52
St. Michael's Long Term Care Centre, 115
Salisbury Greenhouse, 66–67
saskatoons, 39, 41, 47–48, 51–52, 97, 220
Saunders, William, 6, 49–50
schools
    beautification projects, 67, 120, 244, 294n5
    Naturescapes program, 255, 257
    Reeves as children's activist, 67, 244
    tree planting, 120
"The Search for Total Hardiness" (Bugnet), 90
Second World War
    commemorations, 280n45
    vacant lot gardening, 142
    See also victory gardens
seed and plant sources
    Bark Ging Wong's sources, 219–20
    Bugnet's sources, 12, 87–88
    early history, 4, 6, 8, 13, 24–25, 48, 74
    Pike's sources, 13, 76, 80
    Ross's sources, 6–7, 24–25, 50
    seed shortages, 25, 48
    short growing season's impact, 74
    wild *vs.* cultivated distinction, 48
    See also Pike, Alfred
Seed Centre, 66–67, 74, 272n20
Select Roses, Langley, 191
seniors' residences beautification, 114, 115
Seto Deep, 202

Seymour, Patrick, 41
Shewchuk, George
    about, 13–15
    district agriculturalist, 14, 167, 189
    gladioli, 14, 167
    roses, 13–15, 167–69, 189–90
    on Simonet, 97
    writing and teaching, 13–15, 168, 189–91, 287n1
Shoemaker, J.S., 127
shows, exhibitions, and displays
    AHA joint shows, 281n62
    cancellation due to frost, 128
    children in, 244
    city beautification and, 101, 104, 108
    Exhibition Association role, 129, 281n60
    floral displays, 4, 30, 119
    Front Yards in Bloom as replacement for, 135–36
    prizes and trophies, 32, 112, 151–52, 173, 174, 261n32, 281n62
    publicity for, 2–4, 104, 118, 136
    recent trends, 130–31, 135–36, 283n72
    relief gardeners, 151–52, 285n26
    vegetable displays, 3, 3–4, 29–30, 34, 35
    venues, 126, 129, 131
    See also roses, shows
shows, exhibitions, and displays, history
    shows (1879–1886), 2–4, 24–25, 28
    shows (1889–1899), 3, 26, 28–29, 32, 50
    shows (1900–1919), 12, 29–30, 172

shows (1920s), 112, 118, 119, 126,
  127–28, 130–31, 174, 244
shows (1930s), 13, 78, 127–28, 151–52,
  177
shows (1940s), 13, 129, 177, 178–79
shows (1950s), 13, 96, 181–82, 182,
  281n62
shows (1990s), 130
shows (2010 and 2011), 283n72
Shute, Allan, 202–04
Simonet, Lillian, 91, 97
Simonet, Marguerite, 93
Simonet, Robert, 92
  awards, 98
  EHS role, 97
  endowment of U of A award, 98,
    275n75
  life of, 66, 90–92, 97–99
  market gardening, 91–92
  Pike's sales of cultivars, 80–81
  plant breeder, 90–91, 93–94, 96–98,
    187–88, 275n65
  seed and plant sources, 92, 93
Simonet, Robert, plants
  double petunias, xi, 92–93, 98, 187
  fruits and vegetables, 96, 160
  gladioli and hollyhocks, 96
  lilies, xv, 93–94, 95, 98, 264n51
  roses, 94–96, 187–88
Skinner, Frank, 95, 187
social change
  citizen gardeners for, 243–44,
    246–47, 251
  community garden movement,
    163–64

gardening as change agent, 105–06,
  109
Kingwell on, 251, 258
See also schools
soil quality
  community garden movement, 163
  early history, 39–40, 44
  as metaphor, 249
Soldiers of the Soil, 145
Some Native Herbal Remedies (Anderson),
  41
spruce trees, 4, 38, 39, 46, 50, 256
Stevenson, A.P., 275n67
Stiles, H.W., 12, 174, 176
Stowe, Ernest, *184*
  beekeeping, 213
  chief provincial gardener, 211
  city beautification activist, 104,
    276n7
  EHS president, 110, 113
  friendship with Bark Ging Wong,
    211–13, 231
  roses, 172, 176, 177, 179, 183
Stowe, W.A., 139
Strathcona, City of, amalgamation, 111
'Strathcona Gold' apple, 96
Strathcona Horticultural Society
  amalgamation with EHS, 108, 111,
    276n3
  founding of, 102
  shows and exhibits, 12, 29–30, 172
  See also Edmonton Horticultural
    Society
Strathcona Rail Community Garden,
  162

strawberries, cultivated
    experimental farm varieties, 7, 25, 26–27, 50
    named varieties, 7, 26–27, 28
    seed and plant sources, 26, 48
    wild *vs.* cultivated, 26–28, 48
strawberries, wild, 26–28, 47–48, 51
'Strawberry Rhubarb,' 96
Street Railway Barns, 128
suey choy, 219–20, 222, 231
Dr. Sun Yat-Sen Garden, Vancouver, 199
Sustainable Food Edmonton, 135, 141, 164, 254, 287n55
Svedja, Felicitas, 96, 188
sweet peas, 8, 69, 80, 171–72

Tanqueray, John F.D., 268n57
Tarlton, Fred, xv, 94, 264n51
tea roses, 174, 178, 183
Terreault, Oscar, 84
'Thérèse Bugnet' rose
    about, 89–90, 185, 261n28
    Bugnet on, 81–82
    name of, 86
    parentage of, 186–87
    public gardens, 13, 194
thistles, 155, 256–57
Tingley, Ken, 292n11
tobacco, 50
Todd, Frederick G., 53–54, 102, 105
tomatoes, 69, 129, 157, 247, 286n46
trees
    climate and selection, 88
    native species, 38–39, 49–50, 58–59, 123–25, 256
    non-native species, 49–50, 88

    See also Edmonton Tree Planting Committee; fruit trees and bushes
Trees and Shrubs of Alberta (Wilkinson), 38–39
tulips, 67
Turner, George, 38
turnips, 24, 30, 42, 96, 129, 158, 266n15

Ukrainian farmers, 74, 271n19
United States
    Alaska Highway construction, 159
    City Beautiful movement, 103–05
    community gardening, 141
    sales of Simonet's double petunias, 93
University of Alberta, Faculty of Agriculture
    apple breeding, 144, 284n14
    EHS and, 127–28
    founding of, 35
    Pike's partnership with, 77, 80
    research, 77, 80, 127
    rose plots, 127–28
    Simonet's endowments, 98, 275n75
    soil mapping, 39
    support for amateur breeders, 81, 91
    See also Devonian Botanic Garden
University of Alberta, Faculty of Extension, xv, 132
Urban Ag High, 287n55
urban gardening. *See* gardening
Urban Parks Management Plan, 54, 60

vacant lots and community gardening
    about, xiii, 141–43
    balance of protection and
        transformation, 59
    city beautification, 142–43, 147–48
    food security, 144–45, 148, 164
    government policies, 144–49, 164
    recent community garden
        movement, 161–65, 162, 253–54,
        287n55
    statistics on, 131, 142, 143, 146, 151,
        158, 286n46
    See also Sustainable Food
        Edmonton; victory gardens;
        weed control
Vacant Lots Garden Society, 108, 141–
    43, 259n1, 277n15, 277n18
vacant lots program
    amalgamation with Vacant Lots
        Garden Society, 143
    benefits of, 144, 149, 152, 154
    collaboration of EHS and city, 109–
        10, 111–12, 127, 148–51, 277n18,
        278n20, 278n27
    decline of, xiii, 131, 150–51, 156,
        159–61
    educational programs, 151–52,
        285n26
    EHS mandate, 108
    funding and budgets, 112, 156,
        278n20, 278n27
    history, WWI, 108, 112, 141–43,
        277n15
    history, interwar period, 148–51,
        154–56
    history, WWII, 150–51, 156–59

land inventory and rentals, 127, 128,
    146, 161
relief gardeners, 151–52, 285n26
revenue for beautification projects,
    112, 278n20
statistics on, 112, 143, 146, 150–51,
    158, 160, 161, 286n46
victory gardens, 129, 139–41, 151,
    156–59, 286n46
See also weed control
Vavasour, Mervin, 43–44
vegetable gardening. See gardening;
    market gardening
Vick, Roger, 143–44, 275n65
Victoria Avenue (100th Avenue),
    68–69, 69
Victoria Golf Course, 200, 202, 203,
    292n8
Victoria Park Hill, 194
victory gardens, 129, 139–41, 151, 156–
    59, 286n46
violets, 69
Voices of the Soil, 141, 163
Von Aueberg, Paul, 54–56, 60, 267n48
Von Baeyer, Edwinna, 105–06, 132

Wallish Greenhouses, 66–67
Walterdale, 201–02, 217, 292n7
wars. See First World War; Second
    World War
waste places. See vacant lots and
    community gardening
The Way We Green, 269n68
Wayne Wong. See Wong, Ging Wayne

weed control
   collaboration of EHS and city, 111,
      148, 152, 155–56
   early history, 154–55
   educational programs, 155
   inspectors, 155–56
   invasive species, 155–56, 256–57
   recent trends, 256–57
   relief gardeners, 285n26
   vacant lots program, 111, 148, 152,
      154–56, 164, 285n26
Wei Wong. *See* Wong, Ging Wei
West End Park, 55–56
Western Horticultural Society show, 30
What Grows Here (J. Hole), 16
wheat in HBC farms, 24, 42–44, 266n15
Wigelsworth, Harry and Richard,
   272n20
wild plants. *See* native plants
wild roses. *See* roses, wild
wild strawberries. *See* strawberries, wild
Wilkinson, Kathleen, 38–39
willow, 4, 38, 50
Wilson, Alec, 77
Wilson, Walter, *175, 182*
   life of, 176, 183
   publications on roses, 12–13, 78, 80
   "Rose City" campaign, xiii, 12, 78,
      80, 172, 176–79, 181, 185, 194–95,
      289n36
   roses, 13, 177–79, 181–83, *182*
   theatre manager, 12, 78, 176, 194,
      261n32
Wilson, William, 103–05
Winnipeg Horticultural Society show,
   30

Wong, Alfred, 208
Wong, Bark Ging, 204, 207, 212, 214,
   226, 227
   Calder home, 214–15, 217
   cooking and kitchen garden, 218,
      230–31
   Dickinsfield home, 206
   Dovercourt home, 215, 231
   family life, 225–26, 227–28
   friendships, 208, 211–13, 216, 218,
      222, 225, 226, 228, 231
   frugality and generosity, 223–25,
      228–31
   immigration, 206–08, 207, 211, 225
   language learning, 204, 207, 208, 217,
      222, 225, 229
   life in China, 206, 208, 209–10, 213,
      226, 227–28
   marriage and children, 208–9,
      213–14
   off-season life, 229
   retirement, 206, 223, 231
   Stowe's friendship with, 211–13, 231
   traditional culture, 225–27, 229–30
Wong, Bark Ging, market gardening
   early career, 209–11, 215–16
   garden in Calder, 215, 217
   garden in Clover Bar, 215–16
   garden in Dovercourt, 215, 218–19,
      231
   garden in "Namao," 216–24, 231
   garden in Riverdale, *205*, 210–11
   garden near Government House,
      211
   harvest and storage, 223–24
   his views on, 210

hotbeds, 218–19, 221, 231
income and pricing, 216, 223, 224–25
produce, 215, 218–20
seed and plant sources, 219–20
tools and machinery, 210–11, 217–18, 220–22, 224
water, 211, 217, 220–21
wholesale sales, 217–18, 222–24
Wong, Fook Kwan (son), 214, 227
career and education, 206, 215–16, 220, 229, 231–32
life of, 210, 211–14, 216–18, 223–25
Wong, George (Sek), 228
Wong, Ging Wayne (son), 214, 214–18, 223–25, 227, 229, 231
Wong, Ging Wei (son), 214, 227
career and education, 206, 215–16, 222, 229, 231–32, 234
life of, 209, 213–18, 223–25, 231
Wong, Joe (Bark Ging's nephew), 214, 214–18, 223–25, 227, 229
Wong, Lilly Mee (daughter), 213, 215, 231
Wong, Sew and Tong, 231
Wong, Young See (Bark Ging's wife), 205, 212, 214, 226, 227

Calder home, 214–15
cooking and kitchen garden, 218, 230–31
daily routines, 216–18
Dovercourt home, 215
early life in Canada, 205, 211–14
immigration, 211–12
language learning, 212–13, 225
life in China, 208–09, 213, 226, 227
marriage and children, 208–09, 213–14
retirement, 206
See also Wong, Bark Ging
Wong Hop, 203
Wood's rose (*Rosa woodsii*), 89, 186, 287n3
World War One. *See* First World War
World War Two. *See* Second World War
The World We Want (Kingwell), 251, 258
Wright, Percy, 90, 94, 187
Writing Home (Carpenter), 99
Wyatt, F.A., 39

Yard Share, 287n55
Yet You, 203
Young See Wong. *See* Wong, Young See